SUSTAINABLE ENVIRONMENT AND HEALTH

Practical Strategies

SUSTAINABLE ENVIRONMENT AND HEALTH

Practical Strategies

Edited by

Leonardo Sepúlveda Torre, PhD
Ann Rose Abraham, PhD
Neha Kanwar Rawat, PhD
A. K. Haghi, PhD

AAP | APPLE ACADEMIC PRESS

First edition published 2025

Apple Academic Press Inc.
1265 Goldenrod Circle, NE,
Palm Bay, FL 32905 USA

760 Laurentian Drive, Unit 19,
Burlington, ON L7N 0A4, CANADA

CRC Press
2385 NW Executive Center Drive,
Suite 320, Boca Raton FL 33431

4 Park Square, Milton Park,
Abingdon, Oxon, OX14 4RN UK

Library and Archives Canada Cataloguing in Publication

CIP data on file with Canada Library and Archives

Library of Congress Cataloging-in-Publication Data

CIP data on file with US Library of Congress

ISBN: 978-1-77491-650-6 (hbk)
ISBN: 978-1-77491-651-3 (pbk)
ISBN: 978-1-00353-795-3 (ebk)

About the Editors

Leonardo Sepúlveda Torre, PhD
Professor, School of Chemistry, The Universidad Autónoma de Coahuila, Saltillo, Coahuila, México

Leonardo Sepúlveda Torre, PhD, is a full Professor at the School of Chemistry at the Universidad Autónoma de Coahuila, Saltillo, Coahuila, México, where he is also the leader of the Bioprocesses and Microbial Biochemistry Group. His postgraduate studies were related to topics on food biotechnology in the Food Research Department of the UAdeC. In 2011, he conducted research at the Institute of Biotechnology and Bioengineering at the University of Minho (Uminho), Braga, Portugal. He also worked as a collaborator at the Center of Biological Engineering in Uminho, Braga, Portugal, on the project "Biotechnologies for Regional Food Biodiversity in Latin America." He has authored and co-authored 35 scientific articles in indexed journals, several books, and 24 book chapters. Additionally, he holds two national patents.

Ann Rose Abraham, PhD
Assistant Professor, Sacred Heart College (Autonomous), Thevara, Kochi, Ernakulam, Kerala, India

Ann Rose Abraham, PhD, is currently an Assistant Professor at the Department of Physics, Sacred Heart College (Autonomous), Thevara, Kochi, Kerala, India. She has expertise in the fields of condensed matter physics, nanomagnetism, multiferroics, polymeric nanocomposites, etc. She has research experience at various reputed national institutes, including Bose Institute, Kolkata, India; SAHA Institute of Nuclear Physics, Kolkata, India; UGC-DAE CSR Centre, Kolkata, India; and she has collaborated with various international laboratories. She is a recipient of a young researcher award in physics and best paper awards (2020, 2021). She served as Assistant Professor and Examiner at the Department of Basic Sciences, Amal Jyothi College of Engineering, under APJ Abdul Kalam Technological University, Kerala, India. Dr. Abraham is a frequent speaker at national and international conferences. She has a good number of publications to her credit in many peer-reviewed high-impact journals of international repute. She has authored many book chapters and edited more than 10 books with Taylor and Francis, Elsevier, etc.

Neha Kanwar Rawat, PhD
Researcher, Department of Materials Science and Engineering, Indian Institute of Science, Bangalore, Karnataka, India

Neha Kanwar Rawat, PhD, is a recipient of the prestigious DS-Kothari Postdoctoral Fellowship and DST Young Scientist Postdoctoral Fellowship. She is currently a researcher in the Department of Materials Science and Engineering at the Indian Institute of Science in Bengaluru, India. She obtained her PhD in Chemistry from Jamia Millia Islamia (A Central University) in India. Her primary interests lie in nanotechnology, synthesis and characterization of nanostructured materials, with a particular focus on green chemistry. She specializes in sustainable chemical processing of nano-conducting polymers/nanocomposites, conducting films, ceramics, silicones, and matrices such as epoxies, alkyds, and polyurethanes. Additionally, she explores the integration of new technologies in areas such as electrochemistry and organic-inorganic hybrid nanocomposites for biomedical applications. Her research also encompasses protective surface coatings for corrosion inhibition and microwave shielding materials. Dr. Rawat has published numerous peer-reviewed research articles in reputable journals, and her contributions have been featured in book chapters from international publishers.

A. K. Haghi, PhD
Research Associate, Department of Chemistry, University of Coimbra, Portugal

A. K. Haghi, PhD, is a Retired Professor who has written, co-written, edited, or co-edited over 1000 publications, including books, book chapters, and papers in refereed journals. These publications have received over 3800 citations and he has an h-index of 32, according to the Google Scholar database. Currently, he is a research associate at the University of Coimbra, Portugal. Professor Haghi has received several grants and has provided consultation services to major corporations. He is also a frequent speaker at national and international events. He is the Founder and former Editor-in-Chief of the *International Journal of Chemoinformatics and Chemical Engineering* and *Polymers Research Journal.* Additionally, Professor Haghi has served on the editorial boards of many international journals. He has been a member of the Canadian Research and Development Center of Sciences and Cultures. At the University of Guilan (UG), he has supervised several PhD and MSc theses, as well as co-supervised international doctoral projects. Professor Haghi holds a BSc in urban and environmental engineering from the University of North Carolina (USA), as well as two MSc degrees - one in mechanical engineering from North Carolina State University (USA) and another in applied mechanics, acoustics, and materials from the Université de Technologie de Compiègne (France). He was awarded a PhD in engineering sciences from Université de Franche-Comté (France). He is a dedicated reviewer for leading international journals.

Contents

Contributors

Shan Abraham
Post-Graduate and Research Department of Physics, Catholicate College, Pathanamthitta, Kerala, India

Ismael Aguilar-Benitez
EL Colegio de la Frontera Norte (COLEF), Sede Monterrey, Mexico

C. N. Aguilar-González
School of Chemistry, Universidad Autónoma de Coahuila, Blvd. Venustiano Carranza e Ing. José Cárdenas V. s/n. Col., República Oriente, Saltillo, Coahuila, México

Agustín Gómez Álvarez
Department of Chemical Engineering and Metallurgy, University of Sonora, Rosales y Blvd. Luis D. Colosio, Hermosillo, Sonora, Mexico

Juan Alberto Ascacio-Valdés
School of Chemistry, Universidad Autónoma de Coahuila, Blvd. Venustiano Carranza e Ing. José Cárdenas V. s/n. Col., República Oriente, Saltillo, Coahuila, México; Facultad de Ciencias Químicas, Universidad Autónoma de Coahuila, Unidad Saltillo, Saltillo, Coahuila, México

Smriti Bansal
Department of Chemistry, Netaji Subhas University of Technology (N.S.U.T), Azad Hind Fauj Marg, Dwarka, New Delhi, India

Alma C. Campa-Mada
Research Center for Food and Development (CIAD, A.C), Carretera Gustavo Enrique Astiazaran Rosas, Hermosillo, Sonora, Mexico

Elizabeth Carvajal-Millan
Research Center for Food and Development (CIAD, A.C), Carretera Gustavo Enrique Astiazaran Rosas, Hermosillo, Sonora, Mexico

Sergio Colin Castillo
Centro de Investigaciones Socioeconómicas, Universidad Autónoma de Coahuila, Mexico

Agness Chacha
School of Sciences, Jain University, Bangalore, Karnataka, India

Agustín Rascón Chu
Research Center for Food and Development (CIAD, A.C), Carretera Gustavo Enrique Astiazaran Rosas, Hermosillo, Sonora, Mexico

S. C. Esparza-Gonzáles
School of Dentistry-Saltillo, Universidad Autónoma de Coahuila, Avenida Dra, Cuquita Cepeda de Dávila s/n. Col., Adolfo López Mateos, Saltillo, Coahuila, México

Sofía Estrada Flores
Department of Advanced Ceramic Materials and Energy, Facultad de Ciencias Químicas, Universidad Autónoma de Coahuila, Saltillo, Coahuila, México

Adriana Carolina Flores-Gallegos
Facultad de Ciencias Químicas, Universidad Autónoma de Coahuila, Unidad Saltillo, Saltillo, Coahuila, México; School of Chemistry, Universidad Autónoma de Coahuila, Blvd. Venustiano Carranza e Ing. José Cárdenas V. s/n. Col., República Oriente, Saltillo, Coahuila, México

S. D. Flores-Nery
School of Chemistry, Universidad Autónoma de Coahuila, Blvd. Venustiano Carranza e Ing. José Cárdenas V. s/n. Col., República Oriente, Saltillo, Coahuila, México

J. D. García-Ortíz
School of Chemistry, Universidad Autónoma de Coahuila, Blvd. Venustiano Carranza e Ing. José Cárdenas V. s/n. Col., República Oriente, Saltillo, Coahuila, México

José Antonio Rodríguez-de la Garza
Universidad Autónoma de Coahuila, Boulevard Venustiano Carranza, Saltillo, Coahuila, México

Priyanka Ghow
School of Sciences, Jain University, Bangalore, Karnataka, India

Purnima Jain
Department of Chemistry, Netaji Subhas University of Technology (N.S.U.T), Azad Hind Fauj Marg, Dwarka, New Delhi, India

Parvathi Jayasankar
Department of Chemistry and Biochemistry, JAIN (Deemed to be University), Bangalore, Karnataka, India

M. Keerthana
School of Sciences, Jain University, Bangalore, Karnataka, India

K. V. Satheesh Kumar
Postgraduate and Research Department of Chemistry, Government College, Kasaragod, Kerala, India

Antonia Martínez Luévanos
Department of Advanced Ceramic Materials and Energy, Facultad de Ciencias Químicas, Universidad Autónoma de Coahuila, Saltillo, Coahuila, México

Jorge A. Márquez-Escalante
Research Center for Food and Development (CIAD, A.C), Carretera Gustavo Enrique Astiazaran Rosas, Hermosillo, Sonora, Mexico

Silvia Yudith Martínez-Amador
Universidad Autónoma Agraria Antonio Narro, Calzada Antonio Narro, Saltillo, Coahuila, México

U. Meghana
JAIN (Deemed to be University), Bangalore, Karnataka, India

Rekha Mirle
School of Sciences, Jain University, Bangalore, Karnataka, India

Olympia Judith Domínguez Murua
Research Center for Food and Development (CIAD, A.C), Carretera Gustavo Enrique Astiazaran Rosas, Hermosillo, Sonora, Mexico; Department of Chemical Engineering and Metallurgy, University of Sonora, Rosales y Blvd. Luis D. Colosio, Hermosillo, Sonora, Mexico

María Eugenia Molar Orozco
Facultad de Arquitectura de la Universidad Autónoma de Coahuila, Boulevard Fundadores km13, Ciudad Universitaria, Arteaga, Coahuila, Mexico

Deepak Poddar
Department of Chemistry, Netaji Subhas University of Technology (N.S.U.T), Azad Hind Fauj Marg, Dwarka, New Delhi, India

Paul Zavala Rivera
Department of Chemical Engineering and Metallurgy, University of Sonora, Rosales y Blvd. Luis D. Colosio, Hermosillo, Sonora, Mexico

Raúl Rodríguez-Herrera
Facultad de Ciencias Químicas, Universidad Autónoma de Coahuila, Unidad Saltillo, Saltillo, Coahuila, México; School of Chemistry, Universidad Autónoma de Coahuila, Blvd. Venustiano Carranza e Ing. José Cárdenas V. s/n. Col., República Oriente, Saltillo, Coahuila, México

Martin Antonio Encinas Romero
Department of Chemical Engineering and Metallurgy, University of Sonora, Rosales y Blvd. Luis D. Colosio, Hermosillo, Sonora, Mexico

Margarita del Rosario Salazar-Sánchez
Programa de Ingeniería Agroindustrial, Facultad de Ingenierías Universidad del Popular del Cesar, Cesar, Colombia

M. Soosen Samuel
Post-Graduate and Research Department of Physics, Catholicate College, Pathanamthitta, Kerala, India

Steffanny Sanchez-Portillo
Facultad de Ciencias Químicas, Universidad Autónoma de Coahuila, Unidad Saltillo, Saltillo, Coahuila, México

Annie M. Shaji
Post-Graduate and Research Department of Physics, Catholicate College, Pathanamthitta, Kerala, India

Ankita Singh
Department of Chemistry, Netaji Subhas University of Technology (N.S.U.T), Azad Hind Fauj Marg, Dwarka, New Delhi, India

José Fernando Solanilla-Duque
Departamento de Agroindustrial, Facultad de Ciencias Agrarias, Universidad del Cauca, Cauca, Colombia

Binoy T. Thomas
Post-Graduate and Research Department of Botany, Catholicate College, Pathanamthitta, Kerala, India

Johyna Els Thomas
Post-Graduate and Research Department of Physics, Catholicate College, Pathanamthitta, Kerala, India

N. Vedashree
School of Sciences, Jain University, Bangalore, Karnataka, India

V. Vedavalli
JAIN (Deemed to be University), Bangalore, Karnataka, India

Agampodi Sandali Inupama De Zoysa
JAIN (Deemed to be University), Bangalore, Karnataka, India

Preface

Sustainable development, despite being a widely used concept, encompasses various interpretations and elicits diverse responses. In broad terms, sustainable development aims to address growing concerns related to environmental and public health issues, while also considering socioeconomic factors.

The chapters in this new book emphasize the importance of implementing a greener environment, while also taking into account public health, human well-being, and economies. These chapters outline the processes that have been studied, optimized, and developed thus far to ensure the sustainability of our environment and public health.

The book directly presents and examines advancements, developments, and the underlying concepts of creating a healthy urban environment in areas such as water and wastewater treatment, food supply within the context of sustainable development, and chemical contamination. It provides relevant experimental techniques and case studies to explore potential challenges in achieving sustainable practices for urban and environmental communities. This research-oriented volume offers a balanced perspective by including case studies from diverse populations and drawing from the resilience literature.

This book serves as an excellent reference for scholars interested in healthy urban environments. It delves deep into the ethical foundations and sustainable prosperity necessary to achieve sustainable development goals. It also highlights knowledge gaps that can be addressed through novel research in our living environment and future prospects.

CHAPTER 1

OH H_2O! When Resources and Modernism Have No Chemistry

MARÍA EUGENIA MOLAR OROZCO

Facultad de Arquitectura de la Universidad Autónoma de Coahuila, Boulevard Fundadores km13, Ciudad Universitaria, Arteaga, Coahuila, Mexico

ABSTRACT

The aim of this chapter is to highlight the importance of water in our daily life as a vital natural resource for survival of next generation and for a healthy environment. It presents the possibilities to show how to promote the sustainable life and the global sustainable development.

1.1 INTRODUCTION

Let's start with a reflection (AQUIAE, 2021; Conagua, 2019) as in Figure 1.1.

It is widely known that planet Earth is composed of approximately 70% water and the remaining percentage is land surface. Out of the total water, 96.5% is salty and only 3.5% is fresh. Freshwater is distributed among glaciers, soil moisture, aquifers, streams, and rivers, with only 0.025% available to all inhabitants. However, only 0.007% is considered drinkable, although some argue that this percentage may be even lower due to human pollution.

Furthermore, water is one of the vital elements for human life. While a person can survive several weeks without food, they can only last a

Sustainable Environment and Health: Practical Strategies. Leonardo Sepúlveda Torre, Ann Rose Abraham, Neha Kanwar Rawat, & A. K. Haghi (Eds.)

few days without drinking water. Even a mere 10% water deficiency can cause disruptions in the body. The percentage of water in the human body varies according to age and sex, but generally ranges from 39% to 75% (González Pedro & González María José, 2006).

It is worth considering, on the other hand, what Martínez & Patiño (2012, p. 5) point out:

> "Climate and the hydrological cycle are closely linked. Therefore, the increase in temperature and variation in precipitation expected in the most likely climate change scenarios will have a significant impact on the availability of water resources worldwide, including Mexico."

1.1.1 WHAT HAVE WE DONE ABOUT IT?

The actions carried out so far and the ease with which we obtain things have made humanity indifferent and unbelieving in the problem that exists with water. Let's analyze the chain of events:

Accordingly, the UN established 11 objectives, including the sixth objective: Ensure the availability of water, its sustainable management, and sanitation for all.

> "...There are billions of people, mainly in rural areas, who still lack these basic services. Worldwide, one in three people do not have access to safe drinking water, two in five people do not have a basic hand washing facility with soap and water, and more than 673 million people still defecate in the open."

Of the noteworthy goals are:

- "By 2030, the efficient use of water resources in all sectors should be significantly increased, ensuring the sustainability of freshwater abstraction and supply to address water scarcity and reducing the number of people suffering from water deprivation."
- "By 2030, integrated water resources management should be implemented at all levels, including through transboundary cooperation, as appropriate."
- "The participation of local communities in improving water and sanitation management should be supported and strengthened."

Cause	Effect	Consequences
Population growth.	The urban sprawl that absorbs the suburban as well as the rural population is growing.	Concentration in urban localities that directly impacts the use of drinking water, consequently, decreases the average natural availability per capita.
Unsustainable urban development.	The alterations made without good planning in cities, by reducing green areas, as well as aquifers, rivers, together with the increase in impermeable hard surfaces, and water pollution by organic and toxic waste, among others.	Reduction of water catchment, the hydrological cycle is minimized.
The creation of storage dams, the drainage of wetlands, the excessive exploitation of aquifers, etc.	Excessive annihilation of water flows is performed	Reduction of drinking water availability in certain areas or cities, which comes hand in hand with low pressure to supply the most vulnerable population.
Bad habits according to the culture.	Average usage per person per day according to culture (1998 to 2002): • United States 570 liters; • Italy 390 liters; • Japan 380 liters; • Mexico 370 liters; • Denmark 210 liters; • India 140 liters. Less than 50 liters in countries on the poverty line (shortage of liquid).	In places of scarcity, hours are spent collecting and bringing water to homes. This is because water consumption in developed countries decreases water availability in less developed countries.

(Continued Table)

Cause	Effect	Consequences
Anthropogenic activities: Irrigation for agriculture, food production, industry, power generation, domestic use.	A higher percentage is used in agriculture and industry and water is contaminated during the process.	It generates greater demand and reduces the amount of water to be distributed consumption = demand.
The lack of political will in governments to provide universal access to water and sanitation to all inhabitants.	Few resources are invested in budget allocations.	In most cases, those most affected by this exclusion fall on vulnerable areas. Although at present in residential areas they begin to feel the shortage or restrictions.
Climate zone.	Cities located in a greater latitude range have uniform rainfall throughout the year.	Rainfall is important to produce the renewable water of the planet, and this is different according to the region.
Climate change.	It will intensify stress on water resources.	The seasonality of flows in regions will be altered.

Source: Own elaboration; CONAGUA (2019); and Ciudad Redonda (2007).

1.1.2 BUT WHAT WAS IT LIKE BEFORE?

If we look back at the past, it can teach us and sensitize us on this subject. Water was considered sacred in ancient cultures, and therefore, it was cared for and used wisely.

According to Ferro (2006):

> "Water holds strong symbolism and a sacred dimension in cultures across various dimensions and meanings of human life. In all religions and spiritual traditions, water carries a rich meaning that goes beyond its material reality. Water fundamentally symbolizes life. In most creation myths, water represents the source of life and the divine energy that brings fertility to the earth and living beings."

- For Egyptians, water is closely linked to the idea of resuscitation.
- For the Greeks, water has the double meaning of life and death.
- Asian and African people consider fountains as part of their divinities. These fountains are sacred and venerated, serving as a purifying element. One of the most well-known goddesses in Afro religions is Iemanja, who is believed to be the queen and goddess of the waters, representing the divine presence.

For many indigenous people, water is a gift from divinities. It is the abode of spirits. For these people, water is sacred. Many of their traditions affirm in their founding myths that human beings were made or created from water, or that they emerged from water onto the earth in order to take care of nature. Other traditions believe that water is the point of relationship between heaven and earth, the human and the divine (p. 35).

ChaMaqita (2013); and Fayanás (2011):

- Ancient Peru was an agricultural society, and they constructed hydraulic systems to transport water from one location to another, taking into account the force of gravity.
- In the Quechua culture, the community carried out collective work to clean the canals before the distribution of water, and pacts were made with the water owner.
- The Mexica in Mexico is another example, having a relationship with water resources. Their city was designed to operate on water through an aquatic network consisting of chinampas, canals, and ditches.

- The Mayas developed their own technique to increase water pressure. An existing water structure in the Mayan city of Palenque is the oldest known example in America.
- Water is a symbol of birth, rebirth, and fertility for the earth, and is revered in many cultures such as Egypt, Machu Picchu, among others.
- The Egyptians had KRUM, the God of rivers and waterfalls. They believed that the Nile descended from heaven to become a great benefactor.
- The Mochica culture-built pyramids, had a rich agriculture, and had a gigantic network of water channels.

Chávez (2021):

"The valuation of water involves multiple dimensions. In the first instance, there is water for life, which refers to water as a human right. It encompasses the need for everyone to have a daily water supply at home for personal use, food, and hygiene.

Another dimension of valuation is the economic one, where water is primarily seen as a resource or raw material for extractive activities. Unfortunately, mismanagement has led to its overexploitation and pollution, favoring industrial production in sectors such as food, clothing, and technology, both nationally and internationally.

Furthermore, there is the ecological importance of natural water sources. They serve as thermal regulators and provide habitats for endemic species, showcasing their biological uniqueness. These factors justify the need for protection and preservation of these water sources (pp. 1, 2)."

UN (2006):

"Water is also an essential factor in various scientific fields. Each region of the world has its own unique way of consecrating water, but all recognize its value and the central place it occupies in human life. Cultural traditions, indigenous practices, and social values determine how people perceive and manage water resources in different regions of the world (Figure 1.1). However, modern societies are not always aware of the importance of water or what its scarcity means."

FIGURE 1.1 View of the Rhine River in Germany.

Image courtesy: Author

1.1.3 CONSTRUCTIONS CONNECTED TO WATER

The ancient cultures were established near rivers or oceans because they could rely on water to cultivate their crops. Egypt serves as an example:

National Geographic (2016):

> "The Greek historian Herodotus stated that "Egypt is a gift from the Nile." The river, which gives life and serves as an inexhaustible source of resources and communication, had a profound impact on the lives of those who resided along its banks. The Nile River played a fundamental role in the emergence of Egyptian civilization and, with the rhythm of its floods, shaped the lives of its inhabitants. It also played a crucial role in the formation and development of Pharaonic culture (Figure 1.2)."

FIGURE 1.2 Sunrise on the Nile River, Egypt.
Image courtesy: Author

The Egyptians understood the behavior of the river and had a method of predicting its overflow and diverting the waters through canals and dams. They were aware of the impact and knew how to harness it in a natural way.

According to Muñoz & Vidal (2001), the architecture of Angkor showcases the skill of builders, architects, and sculptors in utilizing materials efficiently and intelligently. The highlight, however, lies in its integration with the natural environment and its water architecture. Canals and barayas are seamlessly incorporated into the construction (Figure 1.3). The barayas functioned as artificial structures to retain rainwater, with some reaching lengths of up to 7 kilometers. From these barayas, numerous channels were built to distribute water to the rice fields, palaces, aqueducts, and two moats. Water is a constant presence in this structure, serving not only as a hydraulic system but also as a visual and spatial element (Figure 1.4).

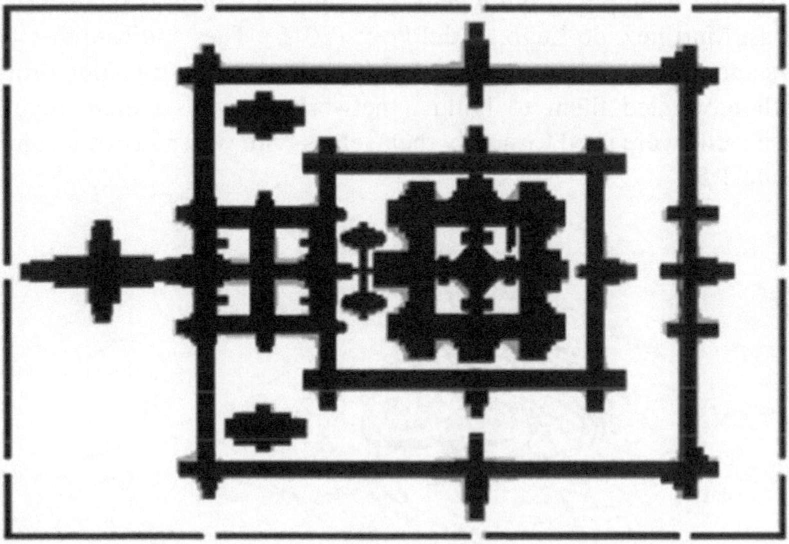

FIGURE 1.3 Architectural plan of Angkor Wat, Cambodia.

Source: Drawn by the author.

FIGURE 1.4 Interior view of Angkor Wat, Cambodia.

Source: Adapted from Roman; Google Earth (2022, 2023).

Another example of the ingenuity of human beings is pointed out by Mejías, Martínez, de Lugo, & del Pozo (2015). They indicate that 4,000 years ago, the inhabitants of La Mancha suffered a long-lasting drought, which motivated them to build a network of wells called montillas. These wells were used to supply themselves with water stored in aquifers (Figure 1.5).

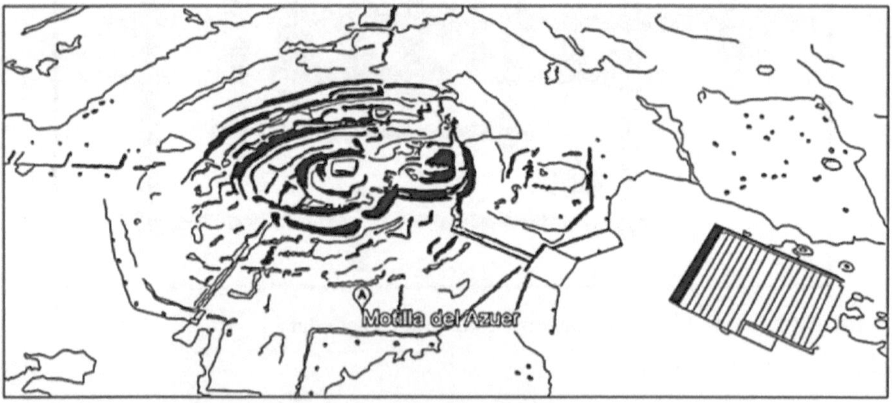

FIGURE 1.5 Aerial view of Montilla de Azuer, Spain.

Source: Adapted from Roman; Google Earth (2018).

Pozzo di San Patrizio is a well located in Orvieto, Umbria, in central Italy. It was built between 1,527 and 1,537 and is an engineering work designed by Antonio Da Sangallo. The well is a cylindrical hollow column (Figure 1.6), situated 63 meters underground and 13 meters wide. It is flanked by two staircases, each consisting of 248 steps. The structure is illuminated by natural light through 72 windows. There are two entrances, one for descending and one for ascending. Even without the help of mules, water could be supplied continuously.

1.1.4 AND WHAT ACTIONS ARE BEING IMPLEMENTED TODAY?

1.1.4.1 FROM THE GOVERNMENT AND INHABITANTS

Currently, there are initiatives and actions being taken by governments to involve the inhabitants. Let's take a look at some examples.

FIGURE 1.6 Exterior view of St. Patrick's Well or Stone Well.
Source: Adapted from Roman; Google Earth (2017).

Since 2000, many governments from different countries have recognized the importance of this issue and have increased investments using national strategies to address it (Ciudad Redonda, 2006).

For instance, India has quadrupled its investment in water supply in vulnerable areas. National budgets play a crucial role in implementing these strategies, along with well-defined goals and political capital that recognize water as a human right.

South Africa has also established this as a basic human right and has demonstrated that it can serve as a mechanism for conferring political power and guidance.

Countries such as Brazil and Thailand, among others, have overcome their technological and economic limitations thanks to innovative national strategies.

In most of these success stories, the achievement has been from the bottom up, with service providers, municipal and local governments working together with practical strategies to solve the problem of access to water. However, they have not been alone. The population with scarce resources has also organized and used their own resources, thus demonstrating that working together and in harmony allows us to achieve goals for the benefit of all.

1.1.4.2 INTERNATIONAL AND NATIONAL ACTIONS

Flores (2021, p. 21) indicates about:

> "The alternative proposed by WWAP (2018, p. 49) was to leverage the wisdom of nature to retain and manage water, both in times of excess and during periods of rainfall and runoff deficit. This is because natural wetlands act as sponges, retaining water and allowing for infiltration. Additionally, unmodified river basins serve as natural reservoirs. This document recommends the implementation of green infrastructure in urban areas, as it plays a crucial role under favorable conditions. Some of the alternatives they suggest are:"

Territorial Scale	Solutions
Basin	• Reforestation and conservation. • Restoration and conservation of wetlands. • Construction of wetlands. • Water harvesting.
Floodplain	• Reconnect rivers to floodplains. • Restoration and conservation of wetlands. • Construction of wetlands. • Water catchment.
City	• Restoration and conservation of wetlands. • Construction of wetlands. • Water catchment. • Green spaces for bioretention and infiltration. • Pavements and permeable streets.

But this requires resources, follow-up, political will over the economy, and personal interests that sometimes limit or tie the hands of actors in Mexico.

Sala, Molinos, Fuentes, & Hernández (2020, p. 202) point out that:

> "In relation to the use of reused water, it is mainly used for agricultural irrigation (70% in the case of Spain). In line with this, the European Union (EU) has recently adopted Regulation (EU) 2020/741 of the European Parliament and of the Council for the reuse of reclaimed water in agricultural irrigation. The regulation aims to ensure that reclaimed water is safe for irrigation, human health, and animal health, promote the circular economy, and contribute to sustainable development. It should not be forgotten that the high cost of treating

and regenerating wastewater acts as a barrier to its reuse. Therefore, the adoption of public policies and tariff systems must consider and value the environmental benefits and other positive externalities derived from water reuse. To achieve this, it is essential to adopt a long-term policy where reused water serves as an additional source to conventional water sources in the water planning of river basins."

Villalba, Murrieta, & Palacios (2021) recommend:

"Urban green areas can make a significant contribution to recharging aquifers due to their ability to infiltrate water. In the case of Xalapa, Veracruz, it is estimated that this contribution could account for 40% of the city's water supply. Vegetation, particularly trees, plays a crucial role in promoting infiltration by protecting the soil and providing organic matter. Additionally, there are artificial alternatives such as porous pavements and infiltration ditches that can be implemented in urban areas and green spaces. These options should be considered in urban planning due to the benefits they offer for water reservoirs, as well as the technical specifications required for their implementation (p. 221)."

Landa, Magaña, & Neri (2008) acknowledge that:

"It requires actions that favor recharging and regulating underground extraction. It is important that the various actions carried out in protected natural areas (ANP), focused on forest management or the maintenance of environmental services, are coordinated with the implementation of strategies to address climate change, in order to enhance them as recharge zones, especially in the semi-arid areas of the central and northern parts of the country. An adequate aquifer management plan would allow for the rotation of extraction based on priorities and rates of overexploitation. This approach requires a detailed analysis to function as an adaptation measure, which in the medium or long term will yield benefits in terms of water availability in aquifers with greater recharge capacity (122 and 123)."

1.1.4.3 ALTERNATIVES IN CONSTRUCTIONS

Baquero (2013, pp. 74–77) explains:

"The consumption of water in a household depends on the number of residents and the types of appliances owned. Therefore, it is crucial

> to consider the implementation of water-saving methods from the initial stages of building design. Wastewater, primarily from domestic sources, can be categorized into sewage (from toilets) and gray water (from showers, sinks, etc.). To promote water conservation, it is recommended to install water-efficient fixtures and appliances, resulting in savings of 25 to 40%.
>
> This system involves connecting the drains of sinks, showers, and bathtubs to a network of pipes that direct the water to a tank. In this tank, the water undergoes treatment and disinfection before being stored for future use, such as filling toilet cisterns (which does not require potable water). Additionally, water from dishwashers and the kitchen can be directed to a grease trap and then to the gray water tank, where it is stored along with water from sinks and showers.
>
> It is important in the network of drainage pipes not to use pipes with a diameter less than four inches due to the possibility of obstructions. The placement of the pipes must have a certain slope, which should not be less than 2%. For the installation of the graywater reuse system, a shut-off valve and a check valve must be implemented, keeping them connected to the main network when it is not sufficient."

By using a graywater reuse system, it represents a productive investment or a value chain by consuming less water from the municipal service. This helps in reducing discharge and treatment costs and contributes to effective sustainability by minimizing stress on aquifers (Baquero, 2013; Antunes & Pasold, 2019).

Sianca, Castro, Him, Gómez, & Velarde (2018) recommend treating washing water for reuse. This benefits the environment by reducing pollution when incorporated. Additionally, they indicate that savings can be achieved in water consumption in homes by reusing it for the next washing process or other non-human consumption uses. The only limitation is the space required for the installation of storage and processing tanks.

Finally, in Monterrey, Nuevo León, Mexico, there was a situation of water scarcity in 2022, which affected all social levels. The government sought alternatives to obtain water from other sources, including cloud seeding to induce rainfall. Alongside these efforts, they also made announcements and publications to encourage residents to be responsible in their use and conservation of water, as depicted in Figures 1.7–1.9.

According to Figures 1.7–1.9, the objective of this work is to describe both the actions and the perception of the inhabitants regarding the subject of water reuse and use.

FIGURE 1.7 A citizen of 100 wash your car with bucket.

Image courtesy: Author

FIGURE 1.8 100 citizens take care of the water.

Image courtesy: Author

FIGURE 1.9 Bath in three minutes.

Image courtesy: Author

1.2 METHODOLOGY

Ethnography is a technique used to understand what users think about various topics. It allows for the measurement, evaluation, and assessment of opinions and attitudes towards specific situations or services within online communities and cultures.

Phases of the method:

1. **Data Collection (Automatic Phase):** Initially, the images are searched and classified. Then, an iconographic analysis is conducted, followed by a study of the context and determination of the message.
2. **Human Interpretation (Manual Phase):** All messages are read, and comments are sorted based on a predefined list of options.

Once all evaluations have been analyzed and a general impression of Internet users' opinions on the subject has been formed, researchers conduct an overall evaluation and interpret the obtained statistics. The aim is to understand users' perspectives on the subject and identify any drawbacks or areas of opportunity.

Based on this classification, the conclusion is formulated.

1.3 RESULTS

The first step is to formulate a phrase or question that enables you to gather relevant information on the topic, in order to narrow down the sources provided by the website. The question used was: "Is humanity committed to preserving water?" The search engine used was Microsoft Bing Images, and the results are as follows:

Approximately 800 images link us to other pages that provide information on actions, care, and awareness made by people, organizations, and institutions on the subject of water. Some of these images may be repeated.

As a next step, the classification was made according to its function, determining that the first 22 images that the search engine retrieved form a collage. When examined together, the search date was found to be January 15, 2023, at 01:35 p.m., sourced from Microsoft Bing Images on the internet.

Now, we will proceed with the analysis of the image content. As a pre-iconographic study, it is observed that 5 images are photographs and 17 are graphics. The predominant colors are blue, green, ocher, black, and red. The photos were taken during the day and depict water, people, water valves, water containers, and messages about the care and importance of water. Iconographically, the main element of the collection is clean and transparent water for human use, depicted through hands delicately treating and caring for it, as well as children consuming it.

In the analysis of the image context, a classification is made based on where they are hosted or sourced:

- News the foreign territory.
- Pinterest and Facebook are partnering with rivers for foreign development.
- Pinterest and Facebook a Green Ecology Association.

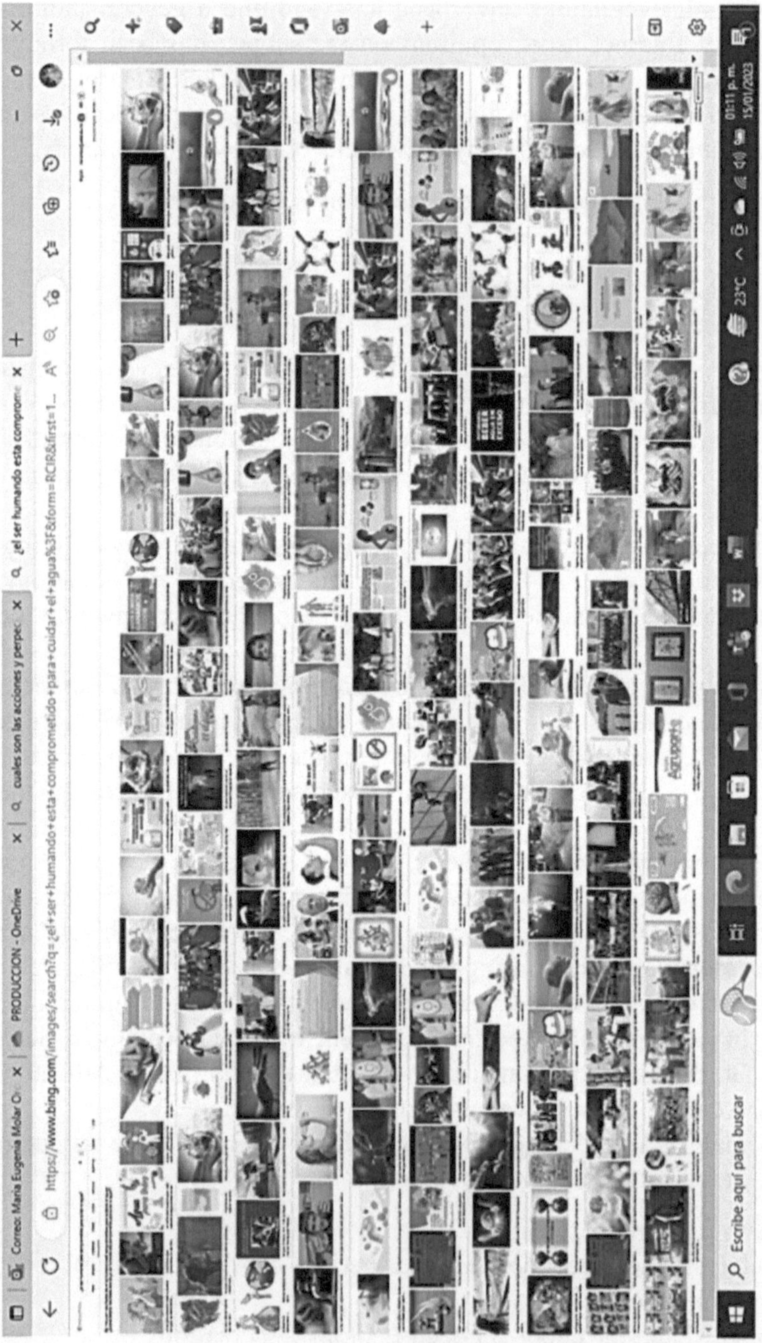
Correo: Maria Eugenia Molar O
PRODUCCION - OneDrive
cuales son las acciones y perpe
¿el ser humando esta comprome
https://www.bing.com/images/search?q=¿el+ser+humando+esta+comprometido+para+cuidar+el+agua%3F&form=RCIRE&first=1...
Escribe aquí para buscar
23°C
01:11 p.m.
15/01/2023

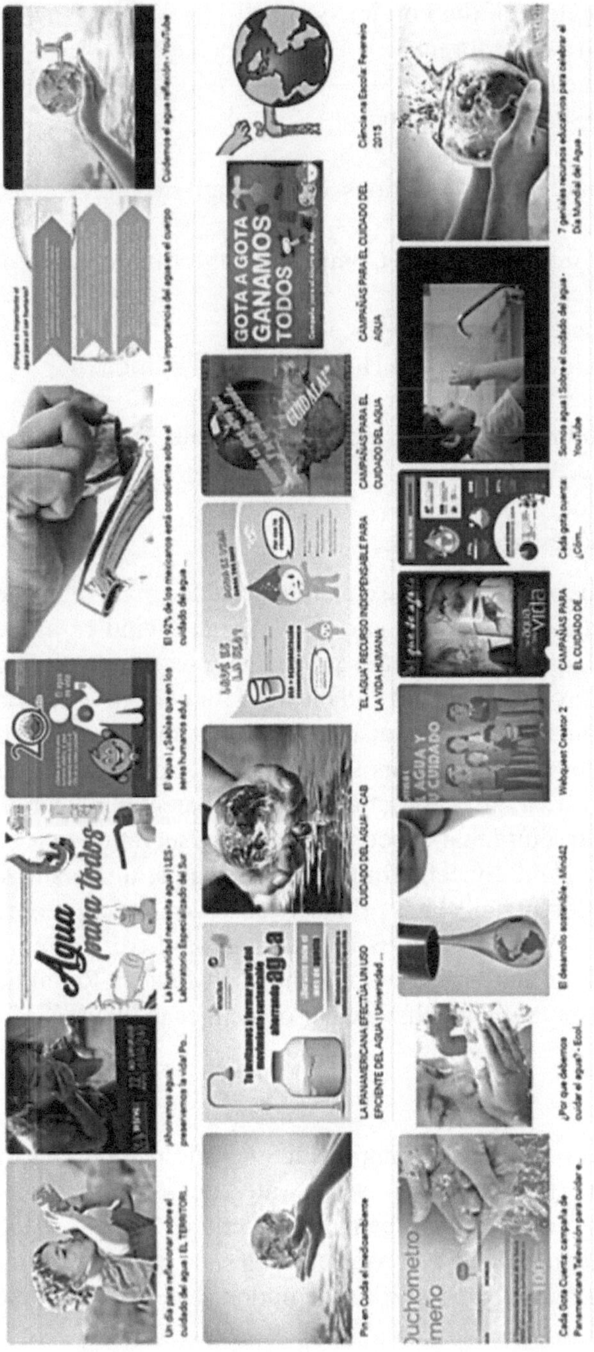
Agua para todos
GOTA A GOTA GANAMOS TODOS
AGUA Y SU CUIDADO

- Specialized laboratory of the Foreign South.
- Flickr CORPOGUAVIO foreign.
- News at the moment, CDMX.
- Plataforma SlideShare from Scribd.
- YouTube (2).
- News of the campaign and actions carried out by Universidad Panamericana (UP).
- Announcements from the Water Company of the Municipality of Boca del Río abroad.
- blogspot.com (there are 5 of this place).
- Panamericana Televisión S.A. 24 Hours Half Day Edition.
- Page taringa.net.
- mind42.com/mindmap program, publicly accessible.
- webquestcreator2.com/
- Noticias The Trade.
- blog.tiching of access to the public.

Based on the previous points, it can be concluded that the main message conveyed is the importance of taking care of water as a vital resource for our survival. The image that internet users perceive when conducting such searches reflects a growing concern and the need for collective efforts to conserve water on our planet. It is also evident that there is limited involvement from the government and its institutions in supporting water conservation, as well as a lack of promotion of environmental education and proposals for infrastructure that encourage water conservation.

Although there is evident concern from various groups, universities, and individuals from different fields, it is primarily observed among people from cultures outside of Mexico. This is concerning, as there are more government-led campaigns and investments within the Mexican Republic that focus on water conservation and have clear objectives in this regard.

1.4 CONCLUSIONS

The objective was achieved by describing the actions and perceptions of the inhabitants regarding the reuse and use of water, demonstrating that this concern still exists among human beings. It is crucial to restore the water cycle and, furthermore, recover wasted water, as well as purify contaminated water and reduce excessive consumption. The key lies in the collective participation through small actions that can lead to significant achievements for the future.

The role of urban planners, government officials, architects, and other professionals is to collaborate in finding better alternatives that promote sustainable development, while also raising awareness among the inhabitants about their role in this endeavor.

The key to driving change and triggering a chain reaction is to reestablish a culture of water, to recognize its vital importance as a non-renewable resource, and to value its conservation. We must all participate with the same goal of sustainability and be aware that our detrimental habits can lead us to a point of no return, where water scarcity becomes a reality.

We must take care of what we love and know, and act accordingly, with the hope that we never have to say:

"Nobody knows what you have until you lose it."

ACKNOWLEDGMENTS

I would like to thank Dr. Leonardo Sepulveda Torre for the invitation to participate in this book at his expense. I would also like to express my gratitude to my husband for inspiring and supporting me in achieving the goals I have set for myself.

KEYWORDS

- **automatic phase**
- **detrimental habits**
- **human interpretation**
- **inhabitants**
- **non-renewable resource**
- **sustainability**
- **urban planners**
- **water scarcity**

REFERENCES

Alfredo, F. M., (2006). The sacred, symbolic, and mystical dimension of water. *Revista Semillas Numero*, (No. 28, 29, pp. 35–38). Retrieved from: https://www.semillas.org.co/apc-aa-files/0ccc57454a31b3c038b4b92d620f7f60/revista-semillas-28-29.pdf (accessed on 24 February 2024).

Antunes De, S., Maria, C. Da. S., & Luiz, P. C., (2019). Water reuse in the field of circular economy and sustainability. *Revista Chilena de Derecho y Ciencia Política, 10*(2), 155–172.

AQUIAE Foundation, (2021). *How Much Drinking Water is There on Earth?* Retrieved from: https://www.fundacionaquae.org/wiki/cantidad-de-agua-potable-fuente-de-vida/ (accessed on 24 February 2024).

AQUIAE Foundation, (sf). *How Much Water is on Earth? And How Much is Suitable for Consumption?* (infographic), Retrieved from: https://www.fundacionaquae.org/cuanta-agua-en-la-tierra/ (accessed on 24 February 2024).

Baquero, M. T., (2013). *Water Saving and Reuse in Buildings in the City of Cuenca, Ecuador* (pp. 71–81). Estoa No. 3.

ChaMaqita, (2013). *Water in Different Cultures*. Rehearsal clubs. Retrieved from: https://www.clubensayos.com/Historia/El-Agua-En-Diferentes-Culturas/681186.html (accessed on 24 February 2024).

CONAGUA, (2019). *Water in the World*. Chapter 8. Retrieved from: https://www.gob.mx/conagua/acciones-y-programas/agua-en-el-mundo (accessed on 24 February 2024).

Edmundo, F. E., (2011). *Pre-Columbian Cultures and Water*. Retrieved from: https://rebelion.org/las-culturas-precolombinas-y-el-agua/ (accessed on 24 February 2024).

FAO (sf), *Agua*. Organización de las Naciones Unidas para la alimentación y la agricultura. Retrieved from: https://www.fao.org/water/es/ (accessed on 24 February 2024).

Flores, E. R., (2021). *Building Urban Resilience in the Face of "Day Zero Water*." Case of Guadalajara, Jalisco, Mexico. Regions and Sustainable Development Magazine, Number 41. Free access.

González, C., & María, J., (2006). *The Water* (Vol. 25, No. 8). OFFAR. Retrieved from: https://www.elsevier.es/es-revista-offarm-4-pdf-13094130 (accessed on 24 February 2024).

González, Pedro (sf), *Qué Porcentaje de Agua Tiene el Cuerpo Humano – Infografía*. AXA Health Keeper. Retrieved from: https://www.axahealthkeeper.com/blog/que-porcentaje-de-agua-tiene-el-cuerpo-humano-infografia/ (accessed on 24 February 2024).

Landa, R., Magaña, V., & Neri, C., (2008). *Water and Climate: Elements for Adaptation to Climate Change* (pp. 1–135). SEMARNAT: Mexico.

Martínez-Austria, P. F., & Patiño-Gómez, C., (2012). Effects of climate change on water availability in Mexico. *Revista Tecnología y Ciencias del Agua, III*(1), 5–20.

Mejias, M. M., Martínez, N., C., Lugo Enrich, Luis B., & Del Pozo, T. J., (2015). Chapter 4. Hydrogeology applied to the study of the motillas of the Bronze Age of La Mancha. Mejías, M. M., Lugo, E., Luis, B., López, S., José, A., & Esteban, L. C., (eds.), *Archaeology, Hydrogeology and the Environment in the Bronze Age: The Culture of Las Motillas*. Madrid: Geological and Mining Institute of Spain.

Muñoz, C. G., & Vidal, L. C., (2001). *Temples of Angkor. More than Five Centuries of History* (No. 11, pp. 96–102). Loggia Architecture & Restoration Magazine. doi: 10.4995/loggia.2001.5234. Retrieved from: https://www.researchgate.net/publication/302455237_Templos_de_Angkor_Mas_de_cinco_siglos_de_historia (accessed on 24 February 2024).

Natalia, C. L., (2021). *The Valuation of Water from Cultural Diversity* (pp. 1–4). IMTA Outlook No. 11. Retrieved from: https://www.imta.gob.mx/gobmx/DOI/perspectivas/2021/b-imta-perspectivas-2021-11.pdf (accessed on 24 February 2024).

National Geographic, (2016). *The Nile, Sacred River of Egypt.* Retrieved from: https://historia.nationalgeographic.com.es/a/nilo-rio-sagrado-egipto_6820 (accessed on 24 February 2024).

Round City, (2007). *Human Development Report 2006.* Retrieved from: https://www.ciudadredonda.org/articulo/informe-sobre-desarrollo-humano-2006#:~:text=Informe%20sobre%20desarrollo%20humano%202006%20El%20saneamiento%20y,humanos%20tienen%20menores%20oportunidades%20de%20desarrollar%20su%20potencial (accessed on 24 February 2024).

Sala-Garrido, R., Molinos-Senante, M., Fuentes, R., & Hernández-Sancho, F., (2020). *Water Reuse: Current Status and Prospects.* Coordinated by the Secretary of State for Budgets and Expenditure. Water in Spain: Economy and governance. Ministry of Finance: Madrid.

Sianca, E., Castro, G., Him, L., Gómez, M., & Velarde, F., (2018). *Treatment and Reuse of Washing Machine Wastewater* (Vol. 4, No. 2, pp. 36–39). RIC.

Sustainable Development Goals (sf), *Goal 6: Ensure the Availability and Sustainable Management of Water and Sanitation for All.* Retrieved from: https://www.un.org/sustainabledevelopment/es/water-and-sanitation/ (accessed on 24 February 2024).

UN, (2006). *UN Highlights Importance of Water in Culture.* Retrieved from: https://news.un.org/es/story/2006/03/1075461 (accessed on 24 February 2024).

Villalba-Murrieta, J. C., Murrieta-Galindo, R., & Palacios-Wassenaar, O., (2021). *Urban Green Areas and Water Infiltration as an Environmental Service in Xalapa* (Vol. 6, No. 1, 2, pp. 202–224). Veracruz, Mexico. Revista Rinderesu.

WWAP, (2018). *United Nations World Water Development Report 2018: Nature-Based Solutions for Water Management.* Paris: UNESCO. Retrieved from: https://unesdoc.unesco.org/ark:/48223/pf0000261579_spa (accessed on 24 February 2024).

CHAPTER 2

Biological Removal of Nitrogen from Wastewater for Sustainable Environment and Health

JOSÉ ANTONIO RODRÍGUEZ-DE LA GARZA[1] and
SILVIA YUDITH MARTÍNEZ-AMADOR[2]

[1]*Universidad Autónoma de Coahuila, Boulevard Venustiano Carranza, Saltillo, Coahuila, México*

[2]*Universidad Autónoma Agraria Antonio Narro, Calzada Antonio Narro, Saltillo, Coahuila, México*

ABSTRACT

The generation of high-nitrogen-content wastewater is mainly due to human activities, such as agriculture (nitrogen fertilizers) and human waste. Since the 1980s, conventional biological treatment processes for the removal of nitrogen present in wastewater have been widely used, such as nitrification (NF), denitrification (DN), dissimilatory nitrate reduction to ammonium (DNRA), and assimilatory nitrate reduction (ANR). Additionally, in recent years, processes such as anaerobic ammonia oxidation (ANAMMOX) and processes derived from this process or the combination of conventional processes such as single reactor system for high ammonia removal over nitrite process (SHARON), completely autotrophic nitrogen removal over nitrite (CANON), oxygen limited autotrophic nitrification denitrification (OLAND), simultaneous partial nitrification, ANAMMOX, and denitrification (SNAD), denitrifying ammonium oxidation (DEAMOX), among others, have captured the attention of researchers in this area since

Sustainable Environment and Health: Practical Strategies. Leonardo Sepúlveda Torre, Ann Rose Abraham, Neha Kanwar Rawat, & A. K. Haghi (Eds.)

treatment costs can be lowered due to the treatment process becoming less energy-intensive (lower dissolved oxygen (DO) requirements, leading to less aeration). Additionally, there are no setbacks during treatment in the cases of wastewater with low organic carbon source concentration (wastewater with low C/N ratio), and the infrastructure space requirements are lesser compared to conventional processes. This chapter aims to give a comprehensive insight into these processes and their implementation in treating high-nitrogen-content wastewater.

2.1 INTRODUCTION

Nitrogen is one of the essential compounds for all forms of life on this planet, and it is a constituent of the main organic macromolecules, such as proteins and nucleic acids, besides being one of the primary nutrients of all living beings. Most of the nitrogen is found as N_2 (gas), part of it is fixed by some living organisms converting this gas into available forms (e.g., NH_4^+), and this process can also be due to anthropogenic causes, for which the balance of the biogeochemical nitrogen cycle is very important (Figure 2.1) (Stevens, 2019; Ward, 2011). Nitrogen is also part of inorganic molecules found in minerals such as sodium nitrate and potassium nitrate salts. Nitrogen is found in various chemical compounds and derived products used daily in agriculture, industry, and at home (Maxwell, 2004). Nitrogen pollution mostly comes from food production due to: (i) excessive use of synthetic nitrogen fertilizers and low absorption of these compounds by plants (low efficiency in nitrogen use); (ii) improper handling of animal waste; (iii) excessive consumption of animal protein (worldwide in 1961 annual consumption was 23.1 kg/person of meat and by 2021 this consumption increased almost double to 42.2 kg/person); and (iv) food waste (Armenta-Bojorquez et al., 2010; Sans & Combris, 2015; Stevens, 2019; Wang et al., 2021). Nitrogenous compounds in wastewater can be removed or transformed using biological processes (Zhu et al., 2008). The processes for biological nitrogen removal (BNR) are divided into conventional ones such as NF, DN, DNRA, and ANR, and advanced processes such as ANAMMOX, SHARON, CANON, OLAND, SNAD, DEAMOX, among others. The application of these processes for the removal of nitrogen depends on factors such as the type of organisms, type and concentration of nitrogenous compounds (which can vary greatly from different sectors such as aniline, chelating agents, corrosion inhibitors (nitrites), dairy, slaughterhouse, and domestic wastes (ammonium, urea, proteins, amino acids)), presence of organic matter, operational conditions, sludge production, energy consumption, and the area required, among other factors (Gerardi, 2002; Magdum & Kalyanraman, 2017; Mulder, 2003; Rosenwinkel et al., 2009; Zhu et al., 2008).

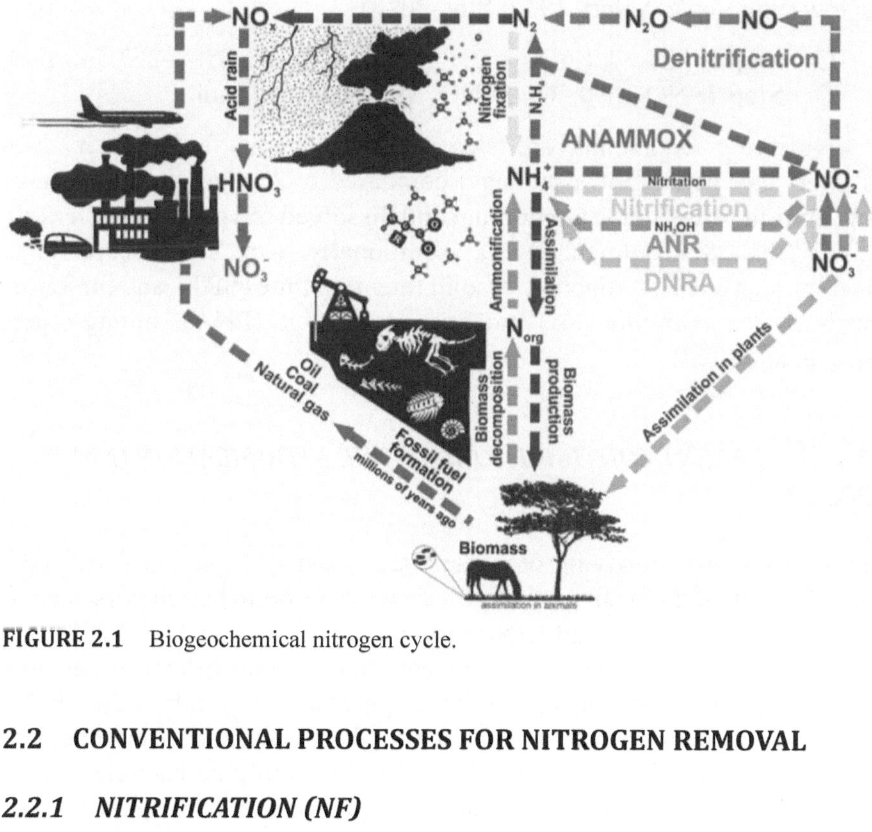

FIGURE 2.1 Biogeochemical nitrogen cycle.

2.2 CONVENTIONAL PROCESSES FOR NITROGEN REMOVAL

2.2.1 NITRIFICATION (NF)

As part of the nitrogen cycle of soils, natural waters, and wastewater treatment systems, nitrification (NF) is an essential process responsible for the biological oxidation of ammonium to nitrate (Ward, 2013). Autotrophic nitrifying organisms carry out NF, with the first step being the oxidation of ammonium to nitrite, and the second step being the oxidation of nitrite to nitrate (Martikainen, 2022). This process is carried out by two groups of bacteria: ammonia-oxidizing bacteria (AOB) or archaea (AOA) catalyze the first step, while nitrite-oxidizing bacteria (NOB) carry out the second step (Alexander, 1965; Costa et al., 2006). While various microorganisms are capable of oxidizing ammonium and nitrite ions, strict aerobic bacteria belonging to the genera Nitrosomonas and Nitrobacter are the main nitrifiers. Nitrosomonas oxidizes ammonium ions to nitrite ions, and Nitrobacter oxidizes nitrite ions to nitrate ions (Gerardi, 2002). The stoichiometric reactions corresponding to the NF process are presented below (Sharma & Ahlert, 1977; Sibi, 2020):

- **Step 1:** $NH_4^+ + 1.5O_2 \rightarrow 2H^+ + H_2O + NO_2^-$ $\Delta G° = -267.5$ kJ/mol.
- **Step 2:** $NO_2^- + 0.5O_2 \rightarrow NO_3^-$ $\Delta G° = -87$ kJ/mol.

Nitrifying organisms are slow-growing (genus Nitrobacter even slower than genus Nitrosomonas) compared to organotrophs and have two limiting substrates: ammonium and dissolved oxygen (DO) (Gerardi, 2002; Shah & Coulman, 1978). Additionally, they are susceptible to factors such as pH, temperature, solid retention time (SRT), and inhibitors such as free ammonia (FA) and free nitrous acid (FNA), among others (Liu et al., 2020).

2.2.1.1 RECENT CONTRIBUTIONS TO THE NITRIFICATION (NF) PROCESS

It is known that nitrifying organisms are autotrophic aerobes with very slow growth and are susceptible to various environmental factors such as DO, temperature, pH, and C/N ratio, among others (Xi et al., 2022). In addition, NF is carried out in two steps by two main genera of bacteria. However, a few years ago, chemolithotrophic bacteria capable of oxidizing ammonium to nitrate in a single step were discovered, a capacity that was previously only known in some fungi and heterotrophic bacteria (Martikainen, 2022). Bacteria possessing heterotrophic nitrifying (HN) and aerobic denitrifying (AD) activity have recently been used; these bacteria are known as HN-AD. HN-AD bacteria have a very versatile metabolism for nitrogen removal that allows them to adapt to the presence or absence of an organic carbon source. This is why they can simultaneously remove high concentrations of nitrogen (ammonium) and carbon at different C/N ratios. It can be said that HN-AD bacteria can oxidize ammonia to hydroxylamine, nitrite, or nitrate and subsequently denitrify to nitrous oxide and nitrogen gas (Xi et al., 2022).

2.2.2 DENITRIFICATION (DISSIMILATORY NITRATE REDUCTION)

The process known as denitrification (DN) is responsible for converting nitrate present in water or soil into reactive nitrogen in the form of N_2 or N_2O. This conversion leads to a loss of nitrogen for the soil or water, as it is not incorporated into cellular material (Firestone, 1982; Gerardi, 2002; Martínez-Espinosa et al., 2021). DN occurs under anoxic conditions and is carried out by facultative anaerobic bacteria, primarily heterotrophic species such as *Alcaligenes, Agrobacterium, Azospirillum, Bacillus, Flavobacterium, Halobacterium, Moraxella, Paracoccus, Propionibacterium, Pseudomonas*, among others (with 80% of nitrate reduction performed by these bacteria, while autotrophs carry out the remaining 20%). Some autotrophic bacteria, like *Thiobacillus,* and photosynthetic bacteria, like *Rhodopseudomonas,* also contribute to this process (André et al., 2011; Firestone, 1982; Knowles, 1982). Heterotrophic denitrifying organisms have two limiting substrates: nitrate and organic carbon (Shah & Coulman, 1978).

In the DN process carried out by heterotrophic bacteria, nitrate is converted to molecular nitrogen through a series of biochemical reactions. This is the general reaction (Sibi, 2022):

$$NO_3^- + 5H_2 + 2H^+ \rightarrow N_2 + 6H_2O \ \Delta G^\circ = -888 \text{ kJ/8e}^- \text{ transfer}$$

Although the energy released can change depending on the substrate, for example, the DN using glucose as a carbon source (Gerardi, 2002):

$$\text{Glucose} + 4.8NO_3^- + 4.8H^+ \rightarrow 6CO_2 + 2.4N_2 + 8.4H_2O + 636 \text{ kcal}$$

In the autotrophic DN, the electron donors are hydrogen (g), reduced sulfur compounds such as sulfur, sulfide, and thiosulfate, trivalent arsenic, arsenite, cyanide, thiocyanate, ferrous iron, and manganese. The inorganic carbon source for this process is carbonate, bicarbonate, or carbon dioxide (Di Capua et al., 2019). Below are some stoichiometric reactions of the autotrophic DN using sulfur compounds as electron donors (Carboni et al., 2021):

$$FeS_2 + 3NO_3^- + 2H_2O \rightarrow Fe(OH)_3 + 2SO_4^{2-} + H^+ \ \Delta G^\circ = -782.3 \text{ kJ/reaction}$$

$$S_2O_3^{2-} + 1.6NO_3^- + 0.2H_2O \rightarrow 2SO_4^{2-} + 0.8N_2 + H^+ \ \Delta G^\circ = -765.7 \text{ kJ/reaction}$$

$$5S^0 + 8NO_3^- + 2H_2O \rightarrow 5SO_4^{2-} + 3N_2 + 4H^+ \ \Delta G^\circ = -547.6 \text{ kJ/reaction}$$

$$5HS^- + 8NO_3^- + 3H^+ \rightarrow 5SO_4^{2-} + 4N_2 + 4H_2O \ \Delta G^\circ = -743.9 \text{ kJ/reaction}$$

2.2.2.1 RECENT CONTRIBUTIONS TO THE DENITRIFICATION (DN) PROCESS

DN is a process carried out by facultative anaerobic microorganisms under anoxic conditions. Recent investigations on aerobic DN have shown that AD bacteria are important in nitrate reduction processes and can survive in a wide range of DO concentrations (3–9 mg/L). Even though aerobic DN is less efficient than anoxic DN, this is attributed to the fact that the bacteria carrying out aerobic DN do not adapt to complex natural environments such as micropolluted water. It has been shown that some actinomycetes, through aerobic DN, can achieve high percentages of total nitrogen and organic carbon removal (Ma et al., 2022).

A recent application of partial denitrification (PDN) (nitrate to nitrite reduction) is its coupling with the ANAMMOX process to replace the partial NF process, which has the disadvantages of unstable nitrite production and requiring aeration. With the combination of the PDN-ANAMMOX processes, a constant supply of nitrite was achieved for the anaerobic oxidation of ammonium with high removal of nitrogen contained in municipal wastewater (Chen et al., 2021).

2.2.3 DISSIMILATORY NITRATE REDUCTION TO AMMONIUM (DNRA)

The DN and DNRA processes are part of the dissimilatory nitrogen cycle. These processes compete for nitrate, and their establishment depends on environmental factors that determine the abundance and conditions favorable for certain microorganisms. DNRA, also known as nitrate ammonification, involves the reduction of nitrate to ammonium, which can remain in the soil unlike N_2. Similar to DN, the nitrite ion serves as an intermediate compound (Liu et al., 2021; Pandey et al., 2020; Rütting et al., 2011; Wang et al., 2020).

The C/N ratio is one of the factors that influence the DNRA and DN processes. DNRA bacteria have an advantage under nitrate-limiting conditions and with a high concentration of electron donors, while denitrifiers are more competitive under nitrate-limiting conditions and electron donor constraints (van den Berg et al., 2017; Wang et al., 2020). These processes occur under anoxic conditions and involve various bacterial genera, including: (i) aerobic bacteria such as *Bacillus, Pseudomonas*, and *Arthrobacter*; (ii) facultative anaerobes such as *Citrobacter*, *Enterobacter*, *Erwinia*, *Escherichia*, *Klebsiella*, *Wolinella*, *Shewanella*, *Photobacterium*, *Caldilinea*, *Aerobacter*, *Marivirga*, *Anaeromyxobacter*, *Vibrio*; and (iii) obligate anaerobes such as *Clostridium*, *Optitus*, *Desulfomonile*, *Ignavibacterium*, *Sulfurospirillum*, *Thiobacillus*, *Desulfovibrio*, *Desulfobulbus*, *Desulfobacterium*, *Thioploca*. It is also known that some fungi (*Ascomycota*) can carry out DNRA, and that certain species of *Paenibacillus* bacteria possess the enzymatic machinery to perform heterotrophic NF, DNRA, and DN. DNRA can be carried out through two metabolic routes: the respiratory route or the fermentative route. In the respiratory route, ATP (energy) is produced through oxidative phosphorylation (electron transport chain), while in the fermentative route, ATP is generated more rapidly through substrate-level phosphorylation (Liu et al., 2021; Pandey et al., 2020; Rütting et al., 2011). The general reaction of DNRA, using nitrate as the final electron acceptor in the oxidation of organic compounds, is presented below (Sibi, 2022):

$$NO_3^- + 4H_2 + 2.4H^+ \rightarrow + NH_4^+ + 3H_2O \ \Delta G° = -603 \text{ kJ/8e}^- \text{ transfer}$$

2.2.3.1 RECENT CONTRIBUTIONS OF THE DISSIMILATORY NITRATE REDUCTION OF AMMONIUM

The knowledge available until a few years ago about the DNRA and DN processes was that they were carried out under anoxic conditions by facultative anaerobic organisms. It has recently been shown that in certain strains of bacteria, such as *Pseudomonas putida*, the processes of ANR, DNRA, and DN can coexist under aerobic conditions, where ANR dominates nitrate reduction for assimilation purposes (Huang et al., 2020).

The nitrate/nitrite-dependent anaerobic oxidation of methane (n-DAMO) process aims to decrease methane emissions into the atmosphere by reducing nitrogenous compounds to reactive nitrogen (ammonium). It was recently detected that in a membrane biofilm reactor (MBfR) fed with methane and nitrate and colonized by n-DAMO archaea and ANAMMOX bacteria, the n-DAMO, DNRA, and ANAMMOX processes can be combined so that the ammonium produced via DNRA is converted to gaseous nitrogen (Nie et al., 2020).

2.2.4 ASSIMILATORY NITRATE REDUCTION (ANR)

By assimilatory nitrate reduction (ANR), only the nitrogen necessary for the biosynthesis and growth of microorganisms is reduced, while in the dissimilatory processes of nitrate reduction (DN and DNRA), nitrate is used as an electron acceptor. The regulator of the ANR process is the concentration of ammonium (what is necessary is produced, if there are surpluses, this process is inhibited); this process is not sensitive to the presence of oxygen, contrary to the dissimilatory processes of nitrate reduction that are insensitive to the presence of ammonium and sensitive to the presence of oxygen (Huang et al., 2020; Tiedje et al., 1981). The assimilatory reduction of nitrate is also a necessary biological process in plants. The nitrate ion is reduced to ammonium to form nitrogenous compounds in the cell. The ammonium ion binds with carbon compounds through two metabolic pathways that convert ammonium into α-amino-nitrogen: glutamate dehydrogenase and glutamate synthetase-glutamate synthase (Guerrero et al., 1981; Guerrero, 1985). Also, some photosynthetic bacteria can reduce nitrate to ammonia for assimilatory purposes (Hunter et al., 2009). However, most microorganisms prefer ammonium ions since nitrate uptake has an energetic cost (Guerrero, 1985; Huang et al., 2020; Sibi, 2022). The reactions corresponding to the ANR are presented below (Guerrero et al., 1981):

- Reduction of nitrate to nitrite in cyanobacteria (ferredoxin-nitrate reduced):

$$NO_3^- + 2Fd_{red} + 2H^+ + 2e \rightarrow NO_2^- + 2Fd_{ox} + H_2O \ \Delta G° = -38.6 \ kcal \cdot mol^{-1}$$

- Reduction of nitrite to ammonium in cyanobacteria (ferredoxin-nitrite reduced):

$$NO_2^- + 6Fd_{red} + 8H^+ + 6e \rightarrow NH_4^+ + 6Fd_{ox} + 2H_2O \ \Delta G° = -103.5 \ kcal \cdot mol^{-1}$$

- Reduction of nitrate to nitrite in fungi, green algae and higher plants (NADPH-nitrate reduced):

$$NO_3^- + NAD(P)H + H^+ + 2e \rightarrow NO_2^- + NAD(P)^+ + H_2O \ \Delta G° = -34 \ kcal \cdot mol^{-I}$$

- Reduction of nitrate to nitrite in fungi, green algae and higher plants (NADPH-nitrate reduced):

$$NO_2^- + 3NAD(P)H + 5H^+ + 6e \rightarrow NH_4^+ + 3NAD(P)^+ + 2H_2O \ \Delta G° = -89.5 \text{ kcal·mol}^{-1}$$

2.3 ADVANCED PROCESSES FOR NITROGEN REMOVAL

2.3.1 ANAEROBIC AMMONIA OXIDATION (ANAMMOX)

The anaerobic ammonia oxidation (ANAMMOX) process is responsible for the biochemical oxidation of ammonia to dinitrogen gas using the nitrite ion as the final acceptor of electrons under anaerobic conditions. This process is carried out by two groups of autotrophic bacteria: the NOB, which partially oxidizes ammonium to nitrite under aerobic conditions, and the ANAMMOX bacteria, which anaerobically oxidize ammonium (Sanabria et al., 2009; Park et al., 2010). More than 50% of oceanic N_2 is produced through the ANAMMOX process, making it a relevant process in the nitrogen cycle (Hong et al., 2009).

The ANAMMOX process has several advantages. It can treat wastewater with high concentrations of ammonium without the need for organic matter. Additionally, the slow growth of ANAMMOX bacteria results in low production of surplus sludge. However, there are also some disadvantages. ANAMMOX bacteria have very slow growth and require at least three months to grow enough ANAMMOX biomass to start a reactor. They are also susceptible to washing or being dragged out of the reactor, and the presence of organic matter can affect their performance (Zhu et al., 2009).

ANAMMOX bacteria are coccoid-shaped and belong to six genera in the phylum *Planctomycetes*. *Brocadia*, *Kuenenia*, *Anammoxoglobus*, *Woebken*, *Jettenia*, and *Scalindua* (Chandel et al., 2022). Currently, various types of bioreactors have been used to couple the ANAMMOX process. In some cases, nitritation occurs first, where a portion of the ammonia ion is converted to nitrite, followed by the ANAMMOX reaction. In other cases, the complete process occurs in a single reactor, such as the CANON and OLAND systems, which are known for their efficiency (Elsayed et al., 2022; Park et al., 2010). The ANAMMOX process is sensitive to the concentration of nitrite and oxygen, so continuous monitoring of these parameters is necessary in the reactor (Park et al., 2010). The stoichiometric reaction of the ANAMMOX process is presented below (Sibi, 2022):

$$NH_4^+ + NO_2^- \rightarrow N_2 + NO_3^- \ \Delta G° = -356 \text{ kJ/mol}$$

2.3.1.1 RECENT CONTRIBUTIONS OF ANAEROBIC AMMONIA OXIDATION (ANAMMOX)

Among the factors that affect the ANAMMOX process is temperature. The optimum range is between 32°C and 37°C, which allows for high nitrogen removal. Conversely, low temperatures result in low activity of the ANAMMOX bacteria (Zhang et al., 2019). This phenomenon has even been observed in an archaeon isolated from Arctic soil, *Candidatus Nitrosocosmicus arcticus*. This archaeon exhibits mixotrophic behavior, being heterotrophic at low temperatures in its organic-rich habitat, and chemolithotrophic in a temperature range of 20°C to 28C, where it oxidizes ammonium. This demonstrates the metabolic versatility of this organism (Alves et al., 2019).

Recently, in an ANAMMOX granular reactor operated at a temperature of 15°C, a nitrogen removal rate of 0.82 kg/(N*m^3*d) was achieved, with a nitrogen load rate (NLR) of 1.23 kg/(N*m^3*d). This was made possible through the acclimatization of microorganisms via intermittent feeding with a high concentration of nitrogen in the influent, along with a gradual decrease in temperature (Zhang et al., 2019).

2.3.2 SINGLE REACTOR SYSTEM FOR HIGH AMMONIA REMOVAL OVER NITRITE

This process can be applied to wastewater with a high ammonium content. It aims to partially oxidize ammonium to nitrite (nitritation), resulting in a mixture containing approximately 50:50 ammonium and nitrite. The effluent from the SHARON reactor can be subjected to other processes, such as heterotrophic DN or ANAMMOX, to convert both nitrogenous compounds (ammonium and nitrite) into gaseous nitrogen (Ahammad & Dolfing, 2020; Hu et al., 2013; Garzón-Zuñiga, 2005; Khin & Annachatre, 2004).

The success of the SHARON process relies on maintaining adequate aeration (3.7 mg/L of dissolved oxygen (DO)), in addition to monitoring the pH (7.0–7.5) and the temperature (30°C–35°C). When combined with the DN or ANAMMOX process, the SHARON process can achieve up to 90% nitrogen removal efficiency, low production of chemical and biological sludge, 25% energy savings during the NF process, and 40% less carbon addition for the DN process (Kalash et al., 2019; Khin & Annachatre, 2004).

An advantage of the ANAMMOX process is its ability to remove a high concentration of ammonium from wastewater that does not contain or contains a very low concentration of organic matter (Garzón-Zuñiga, 2005).

2.3.2.1 RECENT CONTRIBUTIONS OF THE SINGLE REACTOR SYSTEM FOR HIGH AMMONIA REMOVAL OVER NITRATE

The SHARON process was combined with the denitrifying process sulfide removal (DSR) to achieve partial NF (nitrite) through SHARON and a reduction of nitrite to nitrogen gas with the DSR with an influent with a low C/N ratio. A nitrate removal efficiency between 97.7% and 93.5% was detected when the influent feed contained a C/N ratio of 0.646–0.737 in a continuous feeding regime (Guo et al., 2022).

2.3.3 COMPLETELY AUTOTROPHIC NITROGEN REMOVAL OVER NITRITE (CANON)

The completely autotrophic nitrogen removal over nitrite (CANON) process results from the combination of two processes, partial nitritation and ANAMMOX, in a single reactor under limiting oxygen conditions. The CANON process involves the participation of fully autotrophic bacteria, including aerobic bacteria like Nitrosomonas (which convert ammonium to nitrite) and anaerobic Planctomycetes (which convert ammonium and nitrite into nitrogen gas) (Mirbolouki-Tochaei et al., 2019; Sliekers et al., 2002). The CANON process is particularly suitable for treating effluents with high ammonium content and can remove up to 1.5 kg N/m^3 (Tawfik & Elsamadon, 2019). It offers several advantages, such as lower energy requirements for aeration (40%), reduced alkalinity consumption, no need for external carbon or nitrite addition, and minimal sludge production (Gaokar-Desai, 2020; Mirbolouki-Tochaei et al., 2019; Tawfik & Elsamadony, 2019). Moreover, the CANON process is more cost-effective compared to the combined SHARON-ANAMMOX process (Tawfik & Elsamadon, 2019). However, the CANON process has some drawbacks, including the susceptibility of participating microorganisms to FA (concentration below 10 mg/L), free nitrite, and hydroxylamine. It requires proper maintenance of aeration (DO level of 0.29–0.59 g O_2/g N), temperature (between 25°C and 35°C), and sludge retention time (SRT), among other factors (Acton, 2013; Tawfik & Elsamadon, 2019). The CANON process involves the following reactions (Mirbolouki-Tochaei et al., 2019):

$$NH_4^+ + 0.75O_2 + HCO_3^- \rightarrow 0.5NH_4^+ + 0.5NO_2^- + CO_2 + 1.5H_2O$$

$$NH_4^+ + 1.32NO_2^- + 0.066HCO_3^- + 2H^+ \rightarrow 0.066CH_2O_{0.5}N_{0.1} + 1.02N_2 + 0.26NO_3^- + 2.03H_2O$$

$$NH_4^+ + 0.85O_2 \rightarrow 0.44N_2 + 0.11NO_3^- + 1.43H_2O + 0.14H^+$$

2.3.3.1 RECENT CONTRIBUTIONS FROM COMPLETELY AUTOTROPHIC NITROGEN REMOVAL OVER NITRATE

The bacterial composition of the CANON and SNAD processes carried out in suspended sludge and biofilms was studied, as well as the effect of TRH and aeration. These processes were coupled to treat wastewater with a high ammonium content (CANON) and domestic wastewater (SNAD). The results showed that the bacterial composition of the reactors (both CANON and SNAD) with suspended sludge and biofilm reactors contains the same species but in different proportions. The advantages of biofilm reactors include a high proportion and activity of AnAOB (ensuring the ANAMMOX reaction) and denitrifying bacteria (Li et al., 2019). The microbial composition and changes in nitrogen removal were studied in flocs-granules hybrid systems and granule-based systems in CANON reactors. The results showed that the removal of nitrogen, biodiversity, and richness were higher in the hybrid system than in the granule-based system. Likewise, the flocs were responsible for the degradation of ammonium, being important habitats of the AOB and indirectly impacting the role of AnAOBs (Li et al., 2019).

2.3.4 OXYGEN-LIMITED AUTOTROPHIC NITRIFICATION (NF)–DENITRIFICATION (DN)

The OLAND process consists of the combination of two processes: nitritation and ANAMMOX. In this process, ammonium is oxidized by autotrophic bacteria to gaseous nitrogen using nitrite as an electron acceptor under limited oxygen conditions (Vlaeminck et al., 2012). The OLAND process consumes up to 60% less oxygen compared to heterotrophic NF–DN processes, and it does not require the addition of organic matter. Additionally, it generates up to 90% less sludge (Vlaeminck et al., 2012; Windey & Verstraete, 2005).

To achieve optimal performance of the OLAND process, efficient retention of biomass is necessary. It is also important to maintain a balance in the activity of both groups of bacteria (AAOB and AnAOB) to prevent the accumulation of nitrite, as an excess of nitrite can inhibit the activity of the AnAOB (De Clippeleir et al., 2009).

The reactions of the OLAND process (Kuai & Verstraete, 1998) are carried out by two groups of bacteria. The first group is the aerobic ammonium oxidizing bacteria (AAOB), such as Nitrosomonas, which oxidize ammonium to nitrite in the presence of oxygen. The second group is the anaerobic ammonium oxidizing bacterium (AnAOB), such as *Kuenenia sttuttgartensis*, which oxidizes ammonium using nitrite as an electron acceptor (Windey & Verstraete, 2005).

i. $NH_4^+ + 1.5O_2 \rightarrow NO_2^- + H_2O + 2H^+$
ii. $NH_4^+ + NO_2^- \rightarrow N_2 + 2H_2O$

2.3.5 SIMULTANEOUS PARTIAL NITRIFICATION (NF) ANAMMOX AND DENITRIFICATION (DN)

The SNAD process can completely remove organic carbon and ammonium from low-carbon, nitrogen-rich wastewater in a single step (Chacko, 2022; Kumar et al., 2017; Singh et al., 2021). It has been shown that the SNAD process can remove 40 to 50% more total nitrogen than the CANON process (Singh et al., 2021). It is very important to maintain a $COD/NH_4^+ \leq 1$ ratio in the SNAD reactor to avoid inhibition of the ANAMMOX process (Chacko, 2022). The success of the SNAD process will depend on establishing optimal parameters such as pH, temperature, alkalinity, the concentration of limiting substrates (carbon, ammonium, nitrate, and nitrite), and a DO-limited environment. The SNAD process has been successfully established in a variety of bioreactors with different configurations, creating a favorable environment for the various microorganisms involved in the reactions that make up the SNAD process (Kumar et al., 2017). The reactions that occur in the SNAD process can be summarized as follows: (i) ammonium is oxidized to nitrite by AOB (nitrification); (ii) nitrite and ammonium are utilized by ANAMMOX bacteria, forming nitrogen and nitrate; and (iii) nitrate is reduced to nitrogen using organic carbon as an electron donor (Chen et al., 2009; Zhang et al., 2017).

$$2NH_4^+ + 3O_2^- \rightarrow 2NO_2^- + 4H^+ + 2H_2O$$

$$NH_4^+ + 1.32NO_2^- + 0.066HCO_3^- + 0.13H^+ \rightarrow 1.02N_2 + 0.26NO_3^- + 0.066CH_2O_{0.5}N_{0.15} + 2.03H_2O$$

$$NH_4^+ + 0.85O_2^- \rightarrow 0.435N_2 + 0.13NO_3^- + 1.3H_2O + 1.4H^+$$

$$NO_3^- + 1.08CH_3OH + 0.24H_2CO_3 \rightarrow 0.056C_5H_7O_2N + 0.47N_2 + 1.68H_2O + 1.4H^+ + HCO_3^-$$

2.3.5.1 RECENT ADVANCEMENTS IN SIMULTANEOUS PARTIAL NITRIFICATION (NF), ANAMMOX, AND DENITRIFICATION (DN)

The SNAD process was established in an ammonia gas biofiltration system under an autotrophic regime, achieving a high removal of ammonia and total nitrogen. The ammonium was converted to nitrogen gas (97.2–99.5%) and nitrous oxide (0.5–1.8% NT or DN process) at an ammonia loading rate of 0.173 g NH_3/g VSS*d. Ammonia was removed via NF (71%) and ANAMMOX (29%), and nitrogen gas was produced via ANAMMOX (60%) and DN (40%) (Yan et al., 2020).

2.3.6 DENITRIFYING AMMONIUM OXIDATION (DEAMOX)

The denitrifying ammonium oxidation (DEAMOX) process is an alternative for treating wastewater containing ammonium and nitrate (Zhang et al., 2022). This process can be applied to treat nitrogen-rich wastewater that also contains sulfate. In the DEAMOX process, the ANAMMOX and autotrophic denitrifying processes are combined, using sulfur as an electron donor to reduce nitrate ions to nitrite without producing nitrite ions separately and in a controlled manner. The DEAMOX process can achieve a total nitrogen removal efficiency of 90% with a volumetric loading of nitrogen above 1,000 mg N/L/d, but it is important to maintain the proportions of NO_2^-/NH_4^+ (greater than 1.2) and H_2S/NO_3^- (greater than 0.57). One disadvantage of the DEAMOX process is that it requires three reactors: (i) anaerobic; (ii) nitrifying; and (iii) DEAMOX (Chacko, 2022; Kalyuzhnyi et al., 2006; Nazari et al., 2021). The following reactions take place in these reactors (Bassin, 2018):

i. **Anaerobic Reactor:**

$$\text{Organic nitrogen} + SO_4^{2-} \rightarrow NH_4^+ + HCO_3^- + CH_4 + HS^-$$

ii. **Nitrifying Reactor:**

$$NH_4^+ + O_2^- \rightarrow NO_3^- + NO_2^-$$

iii. **DEAMOX Reactor:**

$$4NO_3^- + HS^- \rightarrow 4NO_2^- + SO_4^{2-}$$

$$NH_4^+ + NO_2^- \rightarrow N_2 + 2H_2O$$

2.3.6.1 *RECENT CONTRIBUTIONS OF DENITRIFYING AMMONIUM OXIDATION (DEAMOX)*

The study investigated the effects of adding hydroxylamine and bioaugmenting ANAMMOX bacteria in a DEAMOX reactor. The results demonstrated that the addition of hydroxylamine reduced the start-up period of the reactor by activating bacterial functional metabolism and enhancing partial DN. The DEAMOX system exhibited rapid recovery and maintained excellent performance. Furthermore, nitrate produced by ANAMMOX bacteria could be further reduced to nitrite, indicating its significant role in the microbial community. The combination of hydroxylamine and bioaugmentation resulted in a total nitrogen removal efficiency of 97.22% at a low COD/NO_3^- ratio of 2.6 (Zhang et al., 2022).

2.3.7 *BIOAUGMENTATION FOR NITROGEN REMOVAL PROCESSES*

Bioaugmentation involves the addition of selected strains or mixed cultures to reactors in order to enhance the removal of specific compounds. By introducing bacteria with specific characteristics, bioaugmentation effectively improves reactor performance (Wang et al., 2022). This technique is particularly useful in cases where a large percentage of cell washing is required or when the microorganisms have a slow growth rate (Herrero & Stuckey, 2015).

One example of a bioaugmentation process is BABE (bioaugmentation batch enhanced), which is used to treat nitrogen-rich sidestreams and to enhance cultures of microorganisms involved in NF processes, such as partial nitritation-ANAMMOX, known for their slow growth (Biswas & Nandy, 2015; Chacko, 2022). In the BABE process, the supernatant from the anaerobic digestion of sludge is recirculated to increase the population of nitrifying organisms in the main treatment and facilitate the NF of ammonium from the sidestream (derived from the supernatant stream) (van Haandel & van der Lubbe, 2012).

2.3.7.1 RECENT CONTRIBUTIONS OF BIOAUGMENTATION IN NITROGEN REMOVAL PROCESSES

In a study by Wang et al. (2022), it was found that the bioaugmentation of aerobic granular sludge (AGS) with heterotrophic nitrifiers and aerobic denitrifiers (HN-AD) in a petrochemical industry wastewater treatment reactor resulted in higher efficiency percentages of ammonium and total nitrogen removal compared to control reactors and reactors with free consortium. The reactor with AGS bio-enhanced and immobilized in biocarbon also exhibited a greater abundance of biodegradation of xenobiotic compounds and nitrogen metabolism.

Similarly, in another study conducted by Chen et al. (2022), bioaugmentation with *Acinetobacter* sp. and immobilization of HN-AD bacteria in a moving bed biofilm reactor for swine wastewater treatment significantly improved ammonium and total nitrogen removal. The composition, abundance of bacteria, enzymes, and functional genes related to DN and nitrogen removal showed variations.

2.4 CONCLUSION

A variety of studies on biological nitrogen removal (BNR) processes have been conducted. Recent studies have focused on enhancing the removal of undesirable forms of nitrogen (such as ammonium, nitrogen oxides, nitrite, and nitrate) that contribute to health and environmental issues. Process combinations have been optimized to achieve higher total nitrogen removal, while minimizing sludge production (biomass). These processes operate efficiently without the need for an organic carbon source under low aeration conditions, resulting in lower energy costs and requiring less space for implementation (or combining the processes in a single reactor). Additionally, bioaugmentation and acclimatization techniques can be employed to enhance the removal of various pollutants, not limited to nitrogen-containing wastewater.

The most recent information reveals discoveries about the metabolism of certain microorganisms that eliminate nitrogen through various active metabolic pathways. These include species of the genus Pseudomonas, as well as mixotrophic crops like *Candidatus Nitrosocosmicus.* Additionally, processes such as aerobic DN and heterotrophic NF play a role.

Undoubtedly, understanding the biological removal of nitrogen will continue to provide valuable insights in addressing the problems caused by anthropogenic contamination and the need for remediation.

KEYWORDS

- **ammonium oxidation**
- **anaerobic ammonia oxidation**
- **autotrophic nitrogen**
- **bioaugmentation**
- **denitrification**
- **nitrification**
- **nitrogen removal processes**
- **sustainable environment**
- **wastewater**

REFERENCES

Acton, Q. A., (2013). *Nitrogen Compounds—Advances in Research and Application: 2013 Edition*: Scholarly Brief.

Ahammad, S. Z., Graham, D. W., & Dolfing, J., (2020). Wastewater treatment: Biological. *Managing Water Resources and Hydrological Systems* (pp. 561–576). CRC Press.

Alexander, M., (1965). Nitrification. *Soil Nitrogen, 10*, 307–343.

Alves, R. J. E., Kerou, M., Zappe, A., Bittner, R., Abby, S. S., Schmidt, H. A., Pfeifer, K., & Schleper, C., (2019). Ammonia oxidation by the arctic terrestrial thaumarchaeote

Candidatus nitrosocosmicus arcticus is stimulated by increasing temperatures. *Frontiers in Microbiology, 10,* 1571.

André, L., Pauwels, H., Dictor, M. C., Parmentier, M., & Azaroual, M., (2011). Experiments and numerical modelling of microbially-catalyzed denitrification reactions. *Chemical Geology, 287*(3, 4), 171–181.

Armenta-Bojórquez, A. D., García-Gutiérrez, C., Camacho-Báez, J. R., Apodaca-Sánchez, M. Á., Gerardo-Montoya, L., & Nava-Pérez, E., (2010). Biofertilizantes en el desarrollo agrícola de México. *Ra Ximhai, 6*(1), 51–56.

Bassin, J. P., (2018). New processes for biological nitrogen removal. In: *Advanced Biological Processes for Wastewater Treatment: Emerging, Consolidated Technologies and Introduction to Molecular Techniques* (pp. 143–203). Springer International Publishing AG.

Biswas, R., & Nandy, T., (2015). Biological nitrogen removal in wastewater treatment. In: *Environmental Waste Management* (pp. 95–109). CRC Press Taylor & Francis Group.

Carboni, M. F., Florentino, A. P., Costa, R. B., Zhan, X., & Lens, P. N. L., (2021). Enrichment of autotrophic denitrifiers from anaerobic sludge using sulfurous electron donors. *Frontiers in Microbiology, 12,* 678323.

Chacko, S. P., (2022). Emerging biotechnological processes in controlling nitrogen pollution to minimize eutrophication of surface waters in Asia. In: *Environmental Degradation: Challenges and Strategies for Mitigation* (pp. 125–147). Cham: Springer International Publishing.

Chandel, H., Shyam, K., Kumar, N., Sharma, G., Yadav, M., Murugesan, S., Thakur, S., & Saxena, G., (2022). Anaerobic ammonium oxidation (anammox) technology for nitrogen removal from wastewater: Recent advances and challenges. *Integrated Environmental Technologies for Wastewater Treatment and Sustainable Development* (pp. 23–48). New Delhi, India. Elsevier.

Chen, H., Liu, S., Yang, F., Xue, Y., & Wang, T., (2009). The development of simultaneous partial nitrification, ANAMMOX and denitrification (SNAD) process in a single reactor for nitrogen removal. *Bioresource Technology, 100*(4), 1548–1554.

Chen, H., Tu, Z., Wu, S., Yu, G., Du, C., Wang, H., Yang, E., Zhou, L., Deng, B., Wang, D., & Li, H., (2021). Recent advances in partial denitrification-anaerobic ammonium oxidation process for mainstream municipal wastewater treatment. *Chemosphere, 278,* 130436.

Chen, X., Yuan, C., Zhu, Y., Liu, H., Chen, W., & Zhang, Q., (2022). Bioaugmentation with *Acinetobacter* sp. TAC-1 to enhance nitrogen removal in swine wastewater by moving bed biofilm reactor inoculated with bacteria. *Bioresource Technology, 359,* 127506.

Costa, E., Pérez, J., & Kreft, J. U., (2006). Why is metabolic labor divided in nitrification? *Trends in Microbiology, 14*(5), 213–219.

De Clippeleir, H., Vlaeminck, S. E., Carballa, M., & Verstraete, W., (2009). A low volumetric exchange ratio allows high autotrophic nitrogen removal in a sequencing batch reactor. *Bioresource Technology, 100*(21), 5010–5015.

Di Capua, F., Pirozzi, F., Lens, P. N., & Esposito, G., (2019). Electron donors for autotrophic denitrification. *Chemical Engineering Journal, 362,* 922–937.

Firestone, M. K., (1982). Biological denitrification. *Nitrogen in Agricultural Soils, 22,* 289–326.

Gaokar-Desai, R., (2020). Anaerobic ammonium oxidation (Anammox). In: *Global Implications of the Nitrogen Cycle* (pp. 188–197). Cambridge Scholar Publishing.

Garzón-Zúñiga, M. A., (2005). *Mecanismos no Convencionales de Transformación y Remoción del Nitrógeno en Sistemas de Tratamiento de Aguas Residuales* (Vol. 20, No. 4, pp. 137–149). Ingeniería Hidráulica en México.

Gerardi, M. H., (2003). *Nitrification and Denitrification in the Activated Sludge Process*. John Wiley & Sons.

Guerrero, M. G., (1985). Assimilatory nitrate reduction. In: *Techniques in Bioproductivity and Photosynthesis* (pp. 165–172). Pergamon.

Guerrero, M. G., Vega, J. M., & Losada, M., (1981). The assimilatory nitrate-reducing system and its regulation. *Annual Review of Plant Physiology*, *32*(1), 169–204.

Guo, H., He, T., Chang, J. S., Liu, P., & Lee, D. J., (2022). Nitrogen removal from low C/N wastewater in a novel Sharon & DSR (denitrifying sulfide removal) reactor. *Bioresource Technology*, *362*, 127789.

Herrero, M., & Stuckey, D. C., (2015). Bioaugmentation and its application in wastewater treatment: A review. *Chemosphere*, *140*, 119–128.

Hong, Y., Li, M., & Gu, J., (2009). Bacterial anaerobic ammonia oxidation (anammox) in the marine nitrogen cycle: A review. *Wei Sheng Wu Xue Bao (Acta Microbiologica Sinica)*, *49*(3), 281–286.

Hu, Z., Lotti, T., Van, L. M., & Kartal, B., (2013). Nitrogen removal with the anaerobic ammonium oxidation process. *Biotechnology Letters*, *35*(8), 1145–1154.

Huang, X., Weisener, C. G., Ni, J., He, B., Xie, D., & Li, Z., (2020). Nitrate assimilation, dissimilatory nitrate reduction to ammonium, and denitrification coexist in *Pseudomonas putida* Y-9 under aerobic conditions. *Bioresource Technology*, *312*, 123597.

Hunter, C. N., Daldal, F., Thurnauer, M. C., & Beatty, J. T., (2009). *The Purple Phototrophic Bacteria* (Vol. 28, p. 549). Dordrecht, The Netherlands. Springer.

Kalash, K. R., Kadhom, M. A., & Al-Furaiji, M. H., (2019). Short-cut nitrification of Iraqi municipal wastewater for nitrogen removal in a single reactor. In: *IOP Conference Series: Materials Science and Engineering* (Vol. 518, No. 2, p. 022024). IOP Publishing.

Kalyuzhnyi, S., Gladchenko, M., Mulder, A., & Versprille, B., (2006). DEAMOX—New biological nitrogen removal process based on anaerobic ammonia oxidation coupled to sulphide-driven conversion of nitrate into nitrite. *Water Research*, *40*(19), 3637–3645.

Khin, T., & Annachhatre, A. P., (2004). Novel microbial nitrogen removal processes. *Biotechnology Advances*, *22*(7), 519–532.

Knowles, R., (1982). Denitrification. *Microbiological Reviews*, *46*(1), 43–70.

Kuai, L., & Verstraete, W., (1998). Ammonium removal by the oxygen-limited autotrophic nitrification-denitrification system. *Applied and Environmental Microbiology*, *64*(11), 4500–4506.

Kumar, M., Daverey, A., Gu, J. D., & Lin, J. G., (2017). Anammox processes. In: *Current Developments in Biotechnology and Bioengineering* (pp. 381–407). Elsevier.

Li, B., Wang, Y., Li, X., Zhang, Z., Wang, H., Li, Y., Wu, L., & Li, J., (2019). Comparing the nitrogen removal performance and microbial communities of flocs-granules hybrid and granule-based CANON systems. *Science of the Total Environment*, *703*, 134949.

Li, X., Zhang, J., Zhang, X., Li, J., Liu, F., & Chen, Y., (2019). Start-up and nitrogen removal performance of CANON and SNAD processes in a pilot-scale oxidation ditch reactor. *Process Biochemistry*, *84*, 134–142.

Liu, S., Dai, J., Wei, H., Li, S., Wang, P., Zhu, T., Zhou, J., & Qiu, D., (2021). Dissimilatory nitrate reduction to ammonium (DNRA) and denitrification pathways are leveraged by cyclic AMP receptor protein (CRP) paralogues based on electron donor/acceptor limitation in *Shewanella loihica* PV-4. *Applied and Environmental Microbiology*, *87*(2), e01964–20.

Liu, X., Kim, M., Nakhla, G., Andalib, M., & Fang, Y., (2020). Partial nitrification-reactor configurations, and operational conditions: Performance analysis. *Journal of Environmental Chemical Engineering*, *8*(4), 103984.

Ma, B., Zhang, H., Ma, M., Huang, T., Guo, H., Yang, W., Huang, Y., Liu, X., & Li, H., (2022). Nitrogen removal by two strains of aerobic denitrification actinomycetes: Denitrification capacity, carbon source metabolic ability, and raw water treatment. *Bioresource Technology*, *344*, 126176.

Magdum, S., & Kalyanraman, V., (2017). Existing biological nitrogen removal processes and current scope of advancement. *Research Journal of Chemistry and Environment*, *21*(7), 43–53.

Martikainen, P. J., (2022). Heterotrophic nitrification–An eternal mystery in the nitrogen cycle. *Soil Biology and Biochemistry*, *168*, 108611.

Martínez-Espinosa, C., Sauvage, S., Al Bitar, A., Green, P. A., Vörösmarty, C. J., & Sánchez-Pérez, J. M., (2021). Denitrification in wetlands: A review towards a quantification at global scale. *Science of the Total Environment*, *754*, 142398.

Maxwell, G., (2004). *Synthetic Nitrogen Products: A Practical Guide to the Products and Processes*. Springer Science & Business Media.

Mirbolouki-Tochaei, G., Farrokhi, M., Moslemzadeh, M., Ildari, S., & Mahdavianpour, M., (2019). Conventional and novel biological nitrogen removal processes for nitrogen removal from food industries wastewater. In: *Nitrogen Removal Processes for Wastewater Treatment* (pp. 45–82). Bentham Science Publishers Pte. Ltd. Singapore.

Mulder, A., (2003). The quest for sustainable nitrogen removal technologies. *Water Science and Technology*, *48*(1), 67–75.

Nazari, L., Xu, C., & Ray, M. B., (2021). *Advanced and Emerging Technologies for Resource Recovery from Wastes*. Springer Singapore.

Nie, W. B., Ding, J., Xie, G. J., Yang, L., Peng, L., Tan, X., Liu, B. F., Xing, D. F., Yuan, Z., & Ren, N. Q., (2020). Anaerobic oxidation of methane coupled with dissimilatory nitrate reduction to ammonium fuels anaerobic ammonium oxidation. *Environmental Science & Technology*, *55*(2), 1197–1208.

Park, H., Rosenthal, A., Jezek, R., Ramalingam, K., Fillos, J., & Chandran, K., (2010). Impact of inocula and growth mode on the molecular microbial ecology of anaerobic ammonia oxidation (anammox) bioreactor communities. *Water Research*, *44*(17), 5005–5013.

Rosenwinkel, K. H., Beier, M., Phan, L. C., & Hartwig, P., (2009). Conventional and advanced technologies for biological nitrogen removal in Europe. *Water Practice and Technology*, *4*(1).

Rütting, T., Boeckx, P., Müller, C., & Klemedtsson, L., (2011). Assessment of the importance of dissimilatory nitrate reduction to ammonium for the terrestrial nitrogen cycle. *Biogeosciences*, *8*(7), 1779–1791.

Sanabria, J., Bedoya, L., & Sánchez, J., (2009). Proceso ANAMMOX una aplicación en ingeniería: Revisión general de los aspectos microbianos. *Ingeniería de Recursos Naturales y del Ambiente*, (8), 83–92.

Sans, P., & Combris, P., (2015). World meat consumption patterns: An overview of the last fifty years (1961–2011). *Meat Science*, *109*, 106–111.

Shah, D. B., & Coulman, G. A., (1978). Kinetics of nitrification and denitrification reactions. *Biotechnology and Bioengineering, 20*(1), 43–72.

Sharma, B., & Ahlert, R. C., (1977). Nitrification and nitrogen removal. *Water Research, 11*(10), 897–925.

Sibi, G., (2022). *Environmental Biotechnology: Fundamentals to Modern Techniques*. CRC Press.

Singh, V., Ormeci, B., Mishra, S., & Hussain, A., (2021). Simultaneous partial nitrification, ANAMMOX and denitrification (SNAD)–A review of critical operating parameters and reactor configurations. *Chemical Engineering Journal*, 133677.

Sliekers, A. O., Derwort, N., Gomez, J. C., Strous, M., Kuenen, J. G., & Jetten, M., (2002). Completely autotrophic nitrogen removal over nitrite in one single reactor. *Water Research, 36*(10), 2475–2482.

Stevens, C. J., (2019). Nitrogen in the environment. *Science, 363*(6427), 578–580.

Tawfik, A., & Elsamadony, M., (2019). Completely autotrophic nitrogen removal over nitrite (CANON) process for polishing of anaerobic effluent containing ammonia. In: *Post Treat. Anaerobically Treat. Effluents* (pp. 409–419). IWA Publishing. London, UK.

Van, D. B. E. M., Elisário, M. P., Kuenen, J. G., Kleerebezem, R., & Van, L. M. C., (2017). Fermentative bacteria influence the competition between denitrifiers and DNRA bacteria. *Frontiers in Microbiology, 8*, 1684.

Van, H. A., & Van Der, L. J., (2012). *Handbook Biological Wastewater Treatment-Design and Optimization of Activated Sludge Systems* (2nd edn.). IWA Publishing. London, UK.

Vlaeminck, S. E., De Clippeleir, H., & Verstraete, W., (2012). Microbial resource management of one-stage partial nitritation/anammox. *Microbial Biotechnology, 5*(3), 433–448.

Wang, Q., Kong, J., Liang, J., El-Din, M. G., Zhao, P., Xie, W., & Chen, C., (2022). Nitrogen removal intensification of aerobic granular sludge through bioaugmentation with "heterotrophic nitrification-aerobic denitrification" consortium during petroleum wastewater treatment. *Bioresource Technology, 361*, 127719.

Wang, S., Pi, Y., Song, Y., Jiang, Y., Zhou, L., Liu, W., & Zhu, G., (2020). Hotspot of dissimilatory nitrate reduction to ammonium (DNRA) process in freshwater sediments of riparian zones. *Water Research, 173*, 115539.

Wang, X., Bai, J., Xie, T., Wang, W., Zhang, G., Yin, S., & Wang, D., (2021). Effects of biological nitrification inhibitors on nitrogen use efficiency and greenhouse gas emissions in agricultural soils: A review. *Ecotoxicology and Environmental Safety, 220*, 112338.

Ward, B. B., (2011). Nitrification: An introduction and overview of the state of the field. In: Ward, B. B., Arp, D. J., & Klotz, M. G., (eds.), *Nitrification*.

Ward, B. B., (2013). Nitrification. *Reference Module in Earth Systems and Environmental Sciences*, 1–8.

Windey, K., De Bo, I., & Verstraete, W., (2005). Oxygen-limited autotrophic nitrification–denitrification (OLAND) in a rotating biological contactor treating high-salinity wastewater. *Water Research, 39*(18), 4512–4520.

Xi, H., Zhou, X., Arslan, M., Luo, Z., Wei, J., Wu, Z., & El-Din, M. G., (2022). Heterotrophic nitrification and aerobic denitrification process: Promising but a long way to go in the wastewater treatment. *Science of The Total Environment, 805*, 150212.

Yan, J., Wang, S., Wu, L., Li, S., Li, H., Wang, Y., Wu, J., Zhang, H., & Hong, Y., (2020). Long-term ammonia gas biofiltration through simultaneous nitrification, anammox and denitrification process with limited N_2O emission and negligible leachate production. *Journal of Cleaner Production*, *270*, 122406.

Zhang, B., Zhao, J., Zuo, J., Shi, X., Gong, J., & Ren, H., (2019). Realizing stable operation of anaerobic ammonia oxidation at low temperatures treating low strength synthetic wastewater. *Journal of Environmental Sciences*, *75*, 193–200.

Zhang, F., Peng, Y., Miao, L., Wang, Z., Wang, S., & Li, B., (2017). A novel simultaneous partial nitrification anammox and denitrification (SNAD) with intermittent aeration for cost-effective nitrogen removal from mature landfill leachate. *Chemical Engineering Journal*, *313*, 619–628.

Zhang, X., Zhang, X., Wu, P., Ma, L., Chen, J., Wang, C., Li, X., Liu, W., & Xu, L., (2022). Hydroxylamine metabolism in mainstream denitrifying ammonium oxidation (DEAMOX) process: Achieving fast start-up and robust operation with bio-augmentation assistance under ambient temperature. *Journal of Hazardous Materials*, *421*, 126736.

Zhu, G. L., Hu, Y. Y., & Wang, Q. R., (2009). Nitrogen removal performance of anaerobic ammonia oxidation co-culture immobilized in different gel carriers. *Water Science and Technology*, *59*(12), 2379–2386.

Zhu, G., Peng, Y., Li, B., Guo, J., Yang, Q., & Wang, S., (2008). Biological removal of nitrogen from wastewater. *Reviews of Environmental Contamination and Toxicology*, 159–195.

CHAPTER 3

Vanadium Pentoxide-Based Nanomaterials for the Treatment of Dye-Containing Wastewater to Avoid Negative Effects on the Environment and Human Health

SMRITI BANSAL, ANKITA SINGH, DEEPAK PODDAR, and PURNIMA JAIN

Department of Chemistry, Netaji Subhas University of Technology (N.S.U.T.), Azad Hind Fauj Marg, Dwarka, New Delhi, India

ABSTRACT

Effective wastewater treatment is a key requirement for economic growth in the current water resource-constrained period. The creation and application of cutting-edge, highly effective, and inexpensive wastewater treatment technology is essential. Recent sophisticated techniques in nanomaterial sciences have grabbed scientists' attention in the field of wastewater treatment. However, there is a lack of collective information pertinent to this situation. In this chapter, we will cover the effects of dyes on aquatic vegetation, humans, and water bodies. Dyes hinder photosynthesis, plant growth, bioaccumulation, increase the biochemical and chemical oxygen demand, and possibly promote toxicity, mutagenicity, and carcinogenicity, thereby degrading the quality of water. Furthermore, we compare contemporary technologies frequently employed to remove dyes from wastewater, emphasizing the benefits and shortcomings associated

Sustainable Environment and Health: Practical Strategies. Leonardo Sepúlveda Torre, Ann Rose Abraham, Neha Kanwar Rawat, & A. K. Haghi (Eds.)

with these diverse methods. We also survey prospective advancements in nanotechnology for the treatment of textile dye-containing wastewater. Vanadium pentoxide (V_2O_5)-based nanomaterials, due to their narrow band gap and rapid oxygen transitions, are being employed for the treatment of organic pollutant-containing wastewater, which is discussed in detail in this chapter.

3.1 INTRODUCTION

Water contamination caused by humans and industries does not have a proper disposal system, which is currently one of the world's major problems. Various industries release a huge volume of highly contaminated wastewater containing numerous pollutants. One of the major pollutants is organic dye, which affects human health. Dyes are often classified into different categories based on their origin, structure, and application. The most commonly used synthetic dyes include azo, mordant, acid, basic, sulfide, and many more. Furthermore, textile industries use various types of toxic chemicals or dyes as softening, desizing, finishing, and brightening agents (Affat, 2021). Textile dyes do not bind to fabric and are discarded in their natural form as waste, ending up in aquatic environments such as rivers and ponds, posing toxicological threats to living organisms. Wastewater also contains a wide range of heavy metals and aromatic compounds that are used in the synthesis of dyes (Lellis, Fávaro-Polonio, Pamphile, & Polonio, 2019). These hazardous chemicals persist in water and soil for extended periods of time, causing health risks for living organisms, reducing soil fertility, and slowing down the photosynthetic activity of aquatic species. This results in the development of oxygen-deficient environments for aquatic fauna and flora, decreasing the quality of water bodies by increasing biochemical and chemical oxygen demand (Bashir, Lone, Bhat, Mir, Dar, & Dar, 2020).

Therefore, it is urgently necessary to develop treatment methods that are both economical and environmentally friendly for properly treating wastewater that contains dye before it is finally disposed of. Over the past decades, researchers have tried to solve the aforementioned problems by using various techniques such as membrane filtration, reverse osmosis (RO), photocatalysis, electrocoagulation, chemical precipitation, and many more (Peng & Guo, 2020). Among all these available methods, nano-photocatalysis via Advanced Oxidation Process using semiconductor metal oxide has been widely used due to its various advantages such as energy efficiency, eco-friendliness, and fast pace, as well as its potential to degrade organic dyes using solar light (Chan, Wu, Juan, & Teh, 2011). Vanadium pentoxide (V_2O_5), one of the n-type semiconductors, has a low band gap, enabling it to absorb a broad spectrum of light, resulting in advantages in the field of energy storage, gas detectors, and photocatalysis. Several investigators have used V_2O_5 as a nano-adsorbent and photocatalyst for the degradation of organic dyes under various light sources such as UV-lamp and open environment. To increase the efficiency of V_2O_5, it has been coated with various biopolymers that not only increase its photodegradation percentage but also increase the surface area as well as expedite the adsorption and degradation process (Sajid et al., 2020).

In this chapter, we will provide readers with a comprehensive understanding of the detrimental effects of dye-containing textile wastewater on living organisms and natural ecosystems. Additionally, we will explore various current and cutting-edge treatment methods for effectively managing textile wastewater while promoting environmental safety. Furthermore, we have discussed the impact of chitosan (CH) and CH-anisaldehyde coating on V_2O_5 nanoparticles (CH1, CHA1) for the photodegradation of methylene blue. FTIR, XRD, XPS, and optical band gap analyses were conducted to successfully synthesize the nanoparticles. The results obtained from the synthesized nanoparticles demonstrate that the CH-anisaldehyde coated V_2O_5 nanoparticles, CHA1, exhibited a remarkable 94% photodegradation efficiency in an open environment.

3.2 IMPACT OF TEXTILE DYE ON LIVING BEINGS

Organic dyes are the most abundant pollutants found in the wastewater discharged from textile industries. These dyes possess high thermal and photo-stability, which allows them to persist in the environment for years (Lellis, Fávaro-Polonio, Pamphile, & Polonio, 2019). Organic dyes are highly toxic and carcinogenic, causing significant harm to the environment and leading to various health issues in both humans and animals. The dyes exhibit hazardous effects either in their parent molecule or their intermediate metabolites, such as benzidine, naphthalene, and other aromatic products (Islam, Repon, Islam, Sarwar, & Rahman, 2022). Increased amounts of organic dyes in water bodies disrupt biological processes in aquatic life and hinder the photosynthesis process by blocking sunlight, as well as reducing the water's capacity to reoxygenate.

3.2.1 EFFECT ON HUMAN HEALTH

Various disorders in humans and animals have been associated with textile dyes, as explained in Figure 3.1. These disorders range from dermatitis to issues related to the central nervous system. It is possible that these issues are caused by the inactivation of enzymes due to the substitution of their cofactors. Individuals working in the presence of reactive dyes are susceptible to allergic reactions, including allergic conjunctivitis, contact dermatitis, rhinitis, and occupational asthma (Al-Tohamy et al., 2022). When the reactive dye and conjugate of human serum albumin are generated, acting as an antigen, immunoglobulin E (IgE) antibodies are produced that bind to histamine (Ribeiro & Umbuzeiro, 2014). Additionally, some of the harmful compounds produced by the textile industry may inhibit ovulation and spermatogenesis (Suryavathi et al., 2005). Azo dyes, which are widely used in the paper and textile industries, are synthesized from aromatic amine and its derivatives and have been evaluated for their toxicity (Tounsadi, Metarfi, Taleb, El Rhazi, & Rais, 2020). It has been observed that in mammals, gut microbes metabolize the azo dyes into their parent amine molecules, which are easily absorbed by the intestine. Reactive Green 19 and Reactive Blue 2 are a few examples of textile dyes that have a long-term impact on human health (Leme et al., 2015).

When dyes are used in large quantities, they accumulate in the environment, which affects the food chain of freshwater flora consumed by human beings (Hossain, Das, Islam, Al Mamun, & Khan, 2018). It has been reported that humans have 1,000 times the amount of harmful organic chemicals compared to their initial water concentrations (Korpi, Järnberg, & Pasanen, 2009). Continuous exposure to these substances is detrimental to human health and increases the risk of cancer, as it lowers human immunity (Tounsadi, Metarfi, Taleb, El Rhazi, & Rais, 2020). Azo dyes have the ability to cause mutations. One of the azo dyes, Sudan I, can be enzymatically converted into carcinogenic aromatic amines, which have the capacity to cause cancer (Piątkowska, Jedziniak, Olejnik, Żmudzki, & Posyniak, 2018). Disperse Red 1 dye is mutagenic in nature, as it increases the growth of micronuclei, which is a symptom of mutagenic activity in human lymphocytes and hepatoma (HepG2) cells (*From Target Assessment to Translational Biomarkers*). Disperse Orange 1 dye also harms DNA and induces apoptosis in HepG2 cells by altering base-pair replacement and causing frameshift mutations (Ferraz, Grando, & Oliveira, 2011).

While some dyes, such as Disperse Yellow 7, decompose in natural rivers and produce carcinogenic amines, other dyes, like Direct Blue 14, generate carcinogenic amines when they come into contact with bacterial species present on human skin (Balakrishnan et al., 2016). The metabolite of Methylene Blue, Azure B, is a cationic dye that can intercalate with the helical structures of DNA and duplex RNA. Additionally, this dye has the potential to interact with the lipipoid membrane of animal cells, potentially causing harm by inhibiting the intracellular enzyme monoamine oxidase A, which is responsible for human behavior. Azure B's capacity to block glutathione reductase, an enzyme responsible for maintaining cellular redox balance in the human body, has also been investigated (Li, Zhang, Tang, Zhang, & Mao, 2014). On the other hand, triphenylmethane dyes are harmful to people as they release aromatic amines during decomposition under anaerobic conditions, which can induce skin irritation, allergic dermatitis, mutations, and cancer (Sivarajasekar & Baskar, 2014). Mitotic poisoning from the dye Crystal Violet (triarylmethane) results in chromosomal damage and aberrant metaphase accumulation in various organs, including the vagina and bladder. This carcinogenic compound is also responsible for reticular cell sarcoma (Lellis, Fávaro-Polonio, Pamphile, & Polonio, 2019). Additionally, Crystal Violet has the ability to harm the respiratory and renal systems, as well as the skin, digestive, and immune systems of humans (Mani & Bharagava, 2016).

3.2.2 EFFECT ON AQUATIC LIFE

Aquatic flora is adversely affected when textile dyes are released into aquatic habitats and distributed in various ways, as represented in Figure 3.2. The main issue with organic dye is its reflection and absorption of sunlight, which inhibits its penetration into the photic zone (Liang et al., 2017) of the aquatic environment. This leads to modifications in aquatic habitats and a reduction in photosynthesis in aquatic vegetation (Danish, Abdulateif, Ismail, Sabir, & Zain, 2015). These dyes deteriorate water quality, resulting in dermatitis, allergies, skin irritations, and mutations in humans (Sarvajith, Reddy, & Nancharaiah, 2018). Additionally, an excess accumulation of organic dye reduces oxygen levels, obstructs sunlight, and hinders biological activities (Ghaedi et al., 2015). Moreover, 60%–70% of azo dyes are highly toxic and resistant to traditional treatment methods for physicochemical degradation due to their low biodegradability (Rawat, Sharma, Karmakar, Arora, & Mishra, 2018). If fish and other aquatic species, which are a primary source of protein, accumulate organic dyes and are consumed by humans, it can lead to cramping, fever, and hypertension (Amer, Ali, Azab, El-Shouny, Sun, & Mahmoud, 2022). These organic molecules restrict the water's ability to pass through the fish's gills, blocking gas exchange and potentially slowing down or halting the fish's growth (Berradi et al., 2019). Prolonged exposure to textile effluents also decreases fish feed consumption, resulting in a lower growth rate of macronutrients (Carney et al., 2021).

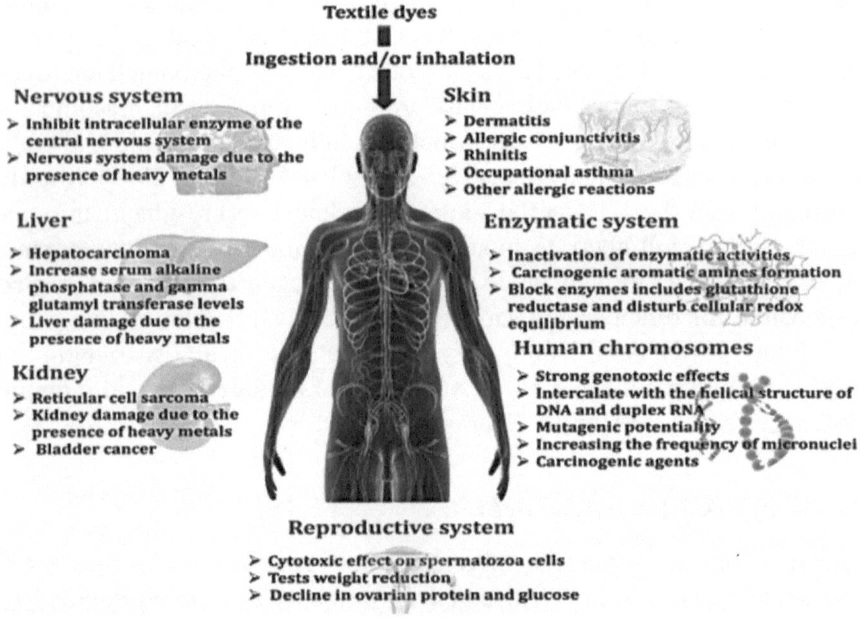

FIGURE 3.1 Effect of organic dyes on human beings.

Source: Reprinted with permission from Al-Tohamy et al., 2022. Copyright © 2022 The Authors. Published by Elsevier Inc

In aquatic habitats, microalgae are the key primary producers that ecologically and economically act as valuable species. However, dye-contaminated water restricts the growth of microalgae and interferes in the transfer of nutrients and energy in aquatic ecosystems. Numerous aspects of algal growth, including pigments, proteins, and other nutrients, are affected by the considerable quantity of dyes released into water bodies. Algae are superior indicators of pollution in toxicological examinations because they are more vulnerable to pollutants than other aquatic species. The cationic dye methylene blue, which is one of the most frequently used dyes in the textile industry, breaks down into aromatic amines like benzidine and methylene, which have the potential to cause cancer. Spirulina platensis and Chlorella vulgaris, two microalgae, were chosen to test the toxic effect of methylene blue (Gelagutashvili, 2013). Methylene Blue dye was exposed to *S. platensis* and *C. vulgaris* at various concentrations, resulting in a decline in the growth rate, pigment content, and protein content. Additionally, Methylene Blue hinders the production of chlorophyll, which restricts their growth and photosynthetic rate.

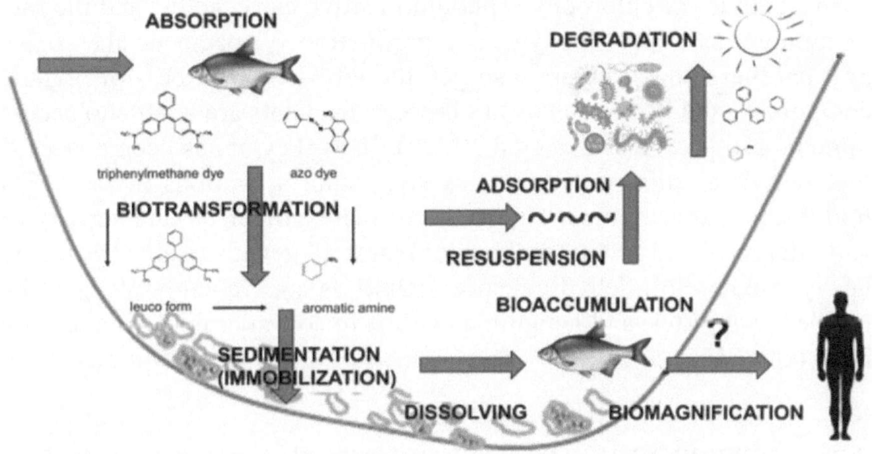

FIGURE 3.2 Different ways in which synthetic organic dyes distributed in aquatic life.

Source: Reprinted from: Tkaczyk, Mitrowska, & Posyniak–Copyright (2020), with permission from Elsevier.

3.2.3 EFFECT ON PLANT GROWTH

The environment for plant growth in developed areas is typically contaminated with a large number of synthetic organic chemicals. Textile dyes are quite stable (environmental half-lives range from 2 to 13 years) and are found in various natural resources, leading to toxicological issues

(Moawad, Abd El-Rahim, & Khalafallah, 2003). Due to population expansion and growing food demand, the use of contaminated streams in agriculture has expanded in many nations. However, frequent irrigation of agricultural regions with contaminated water from various sectors can harm the fertility of the soil as well as the growth of plants (Field, Campbell, & Lobell, 2008). Studies have shown that plants exposed to effluents from the textile industry had lower levels of carbohydrates, proteins, and chlorophyll (Kaushik, Garg, & Singh, 2005). One researcher reported the toxic effects of textile mill effluents on two types of sorghum, "Pioneer jowar" and "Desi jowar." Due to the presence of effluents, seed germination, plant shoot and root length, biomass, and levels of carotenoids and chlorophyll decrease.

Four azo dyes and two anthraquinone dyes were applied to foliage at varying doses. They had an adverse effect on the pace of photosynthesis, amount of chlorophyll, photoprotective carotenoids, and the rate of monoterpene and lipoxygenase production (Copaciu et al., 2013). Soil hardness and structure also get altered when an excess of organic compounds and heavy metals are present, as plants are unable to access ionic nutrients (Srivastava et al., 2020). In Turkey, it has been observed that industrial wastewater induces significant alterations in the soil's biological characteristics. The health of plants, diversity and activity of soil microbes, and genetic makeup of plants all get adversely affected by heavy metals. Soil dehydrogenase activity is consequently elevated by textile wastewater, and a significant shift in soil microbial biomass has been noticed.

3.3 DIFFERENT TYPE OF TECHNOLOGY AND METHODOLOGY USED FOR THE TREATMENT

Discharged textile dyes persist in the environment for extended periods of time, necessitating the development of a sustainable procedure for the degradation of organic dyes. In terms of legislation, there is currently no worldwide agreement on the discharge of wastewater. Depending on the impurities present in the wastewater, physical, chemical, biological, or a combination of these methods can be adopted for treatment, as mentioned in Figure 3.3. Some of the most prevalent approaches are as in subsections.

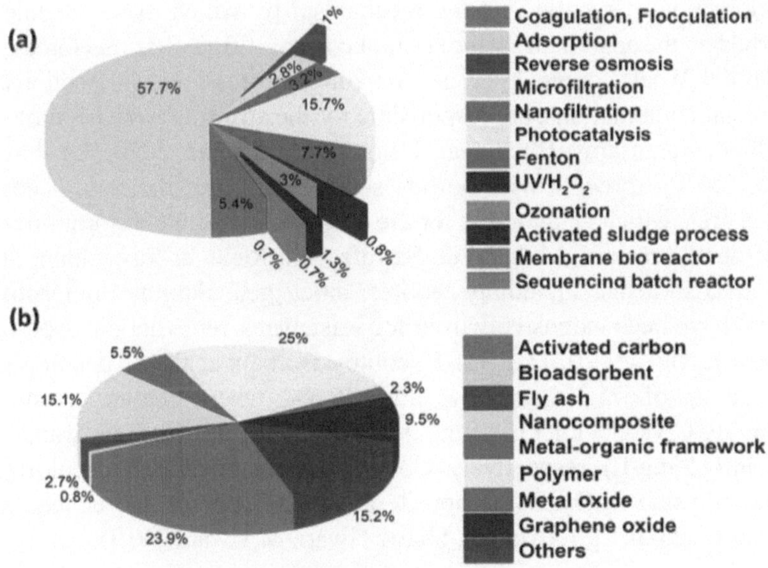

FIGURE 3.3 Pie-chart representing various physical, chemical, and biological methods for the treatment of dye-containing water.

Source: Reprinted with permission from Dutta et al., 2021. https://creativecommons.org/licenses/by/3.0/

3.3.1 PHYSICAL TECHNIQUES

Different physical techniques, such as ion exchange, adsorption, and membrane filtration, have been utilized for the removal of dye from wastewater based on the mass transfer mechanism. These techniques possess high removal efficiency, ranging from 85% to 99% (Al-Tohamy et al., 2022). There are numerous benefits associated with physical approaches, including simple designs, ease of use, low costs, limited chemical requirements, and the absence of harmful compounds with no inhibiting effect (Cao et al., 2021). However, there are also several disadvantages, such as the production of poisonous byproducts and restricted usage, which generally make these technologies unfavorable (Akpomie & Conradie, 2020).

3.3.1.1 ADSORPTION PROCESS

It is a process in which ions or molecules are adsorbed on the surface of a solid adsorbent. There are two forms of adsorption: physisorption and chemisorption, depending on the mechanism by which dye molecules are adsorbed on the adsorbent surface (Burakov et al., 2018). Various forces, such as van der Waals forces, hydrogen bonding, hydrophobic interactions, and electrostatic interactions, are responsible for the adsorption of dye molecules (Mudhoo, Ramasamy, Bhatnagar, Usman, & Sillanpää, 2020), as shown in Figure 3.4. Adsorbents with porous surfaces enhance the total accessible surface area, which is favorable for the quick and effective adsorption of dye molecules (Samsami, Mohamadi, Sarrafzadeh, Rene, & Firoozbahr, 2020). Several adsorbents, including zeolites, silica gel, alumina, and activated carbon, have been extensively used for wastewater remediation (Abu-Nada, Abdala, & McKay, 2021). ZSM-5 zeolite adsorbent achieved decolorization percentages of 75.3%, 81.2%, and 86.6% when treating wastewater containing Crystal Violet (15 mg/L), Methylene Blue (15 mg/L), and Basic Fuchsin (15 mg/L), respectively. Congo Red exhibited 90% decolorization efficiency using ZnO@Ze composite as an adsorbent in the concentration range of 0.025–0.1 g/L (Madan, Shaw, Tiwari, & Tiwari, 2019).

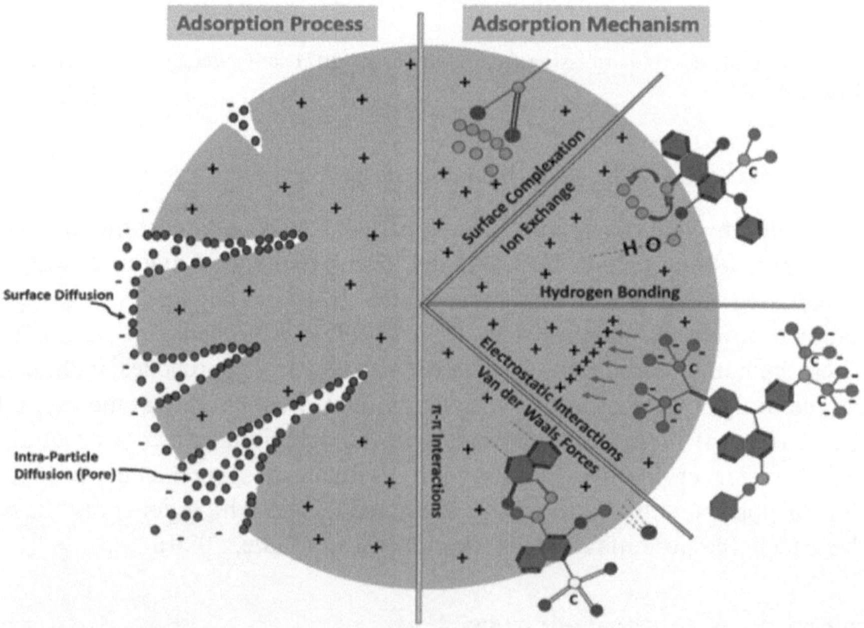

FIGURE 3.4 Adsorption mechanism and process for degradation of dye.

Source: Reprinted with permission from Hamad et al., 2022. Copyright © 2022 by the authors. Licensee MDPI, Basel, Switzerland. https://creativecommons.org/licenses/by/4.0/

3.3.1.2 ION-EXCHANGE METHOD

This method has suddenly attracted a lot of attention in the treatment of wastewater due to its benefits, which include regeneration, simplicity, adaptability, affordability, and high efficiency. The bonds between the resins and the solutes are responsible for facilitating effective separation in this method (Figure 3.5) (Akpomie & Conradie, 2020; Ahmad et al., 2015). One of the cation exchangers removes 91.7% of Disperse Violet 28, while Amberlite IRA 400, an anion-exchange resin, removes 97% of Acid Orange 10 (Bayramoglu, Kunduzcu, & Arica, 2020).

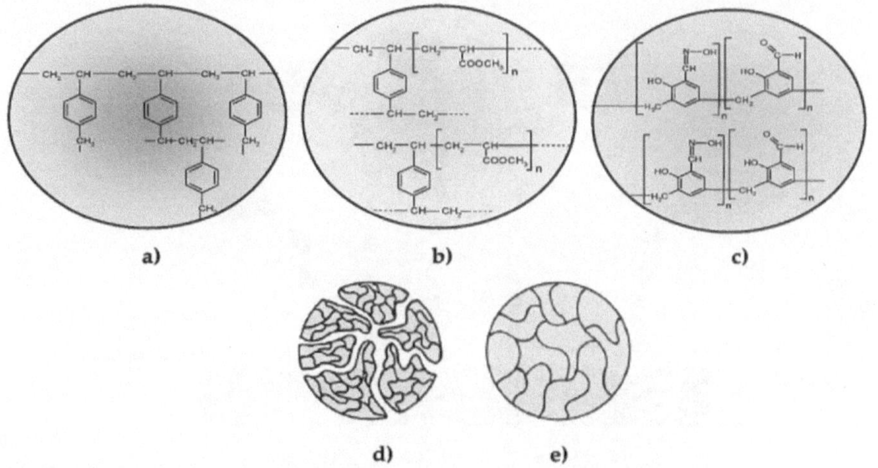

FIGURE 3.5 Different ion-exchangers involved in dye degradation process.

3.3.1.3 MEMBRANE FILTRATION

One of the most advanced physical technologies for dye removal is membrane filtration. Membrane filtration works by trapping solutes larger than the pores of the membrane, resulting in dye-free water. Microfiltration (MF) is a membrane-based filtration method that uses a standard membrane with pores ranging in size from 0.1 to 10 μm to remove dyes from wastewater. Nanofiltration, on the other hand, utilizes membranes with a diameter between 0.5 and 0.2 nm and has recently been applied for the treatment of dye-containing wastewater (Behera, Nayak, Banerjee, Chakrabortty, & Tripathy, 2021). The use of size and electrostatic repulsion principles in nanofiltration technology allows for the separation of dye molecules from wastewater solutions (Dasgupta et al., 2015).

Ultrafiltration membranes, with widths ranging from 0.1 to 0.001 μm, can also be used to extract organic pigments from textile wastewater. Although cheaper than nanofiltration, ultrafiltration has a lower separation rate due to its larger membrane pore size (Dasgupta et al., 2015). RO, a widely used membrane filtering technology in the industry, is effective in cleaning wastewater containing dyes and producing high-quality water (Pei, Wang, You, Li, & Ren, 2022) (Figure 3.6).

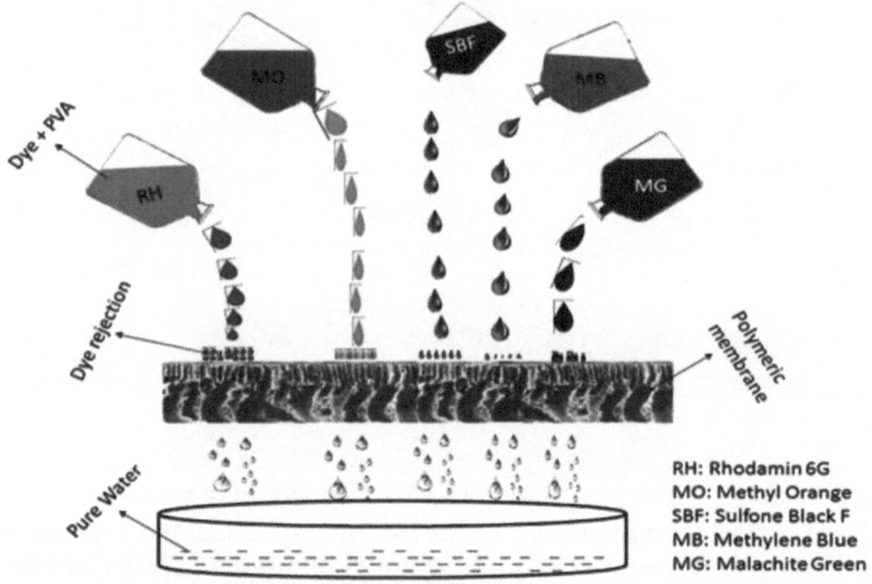

FIGURE 3.6 Membrane filtration to get pure water.

Source: Reprinted from: Sunil et al.–Copyright (2021), with permission from Elsevier.

3.3.2 CHEMICAL METHODS

Dye-containing wastewater can be treated chemically using techniques such as coagulation, electrochemistry, and advanced oxidation procedures. However, these techniques are often more expensive than other methods. Furthermore, the need for appropriate equipment, high electrical energy requirements, and large quantities of chemicals used are the main disadvantages of using chemical methods to remove dye from textile effluent (Kishor et al., 2021). Another limitation of this technique is the production of hazardous metabolites and by-products during the treatment process (Katheresan, Kansedo, & Lau, 2018).

3.3.2.1 COAGULATION–FLOCCULATION

In water treatment, flocculation and coagulation are employed to remove suspended solids by destabilizing the suspended particles in wastewater. Polymers and metal salts can be used as coagulants in this treatment method, while polymers are used as flocculants to enhance the aggregation of flocs for easier separation (Al-Mutairi, 2006), as shown in Figure 3.7. During the vigorous mixing step, coagulants are introduced to neutralize or reduce the charge of finely distributed particles (Zarei et al., 2018). Various coagulants such as magnesium carbonate, hydrated lime, ferrous sulfate, and ferric chloride sulfate are used to maintain purity in the system (Li et al., 2018). When a coagulant is added to a solution, it precipitates and traps organic pollutants and impurities. The precipitated substances can then be physically filtered to produce uncontaminated water. Acid Orange 7 showed a decline of 98% when laterite soil, rich in iron and aluminum, was used as a flocculant and coagulant. This method is cost-effective; however, it has drawbacks such as the production of massive sludge and pH-dependent separation.

3.3.2.2 ADVANCED OXIDATION PROCESSES

Advanced oxidation techniques include photocatalysis, Fenton, photo-Fenton, ozonation, and electrochemical procedures as mentioned in Figure 3.8. These techniques effectively remove dye from challenging environments without producing sludge. However, they have several drawbacks such as high cost, pH dependence, and the formation of hazardous by-products (Zhao, Cao, Lu, Li, Pang, & Wu, 2022). In order to generate free radicals and holes for dye degradation via photocatalysis, nanoparticles like V_2O_5 and titanium peroxide are utilized as photocatalytic agents (Siwińska-Stefańska et al., 2018). The holes oxidize organic dye molecules, producing $OH^\bullet$, while the electrons generated by superoxide act as reducing agents (Hassanshahi & Karimi-Jashni, 2018).

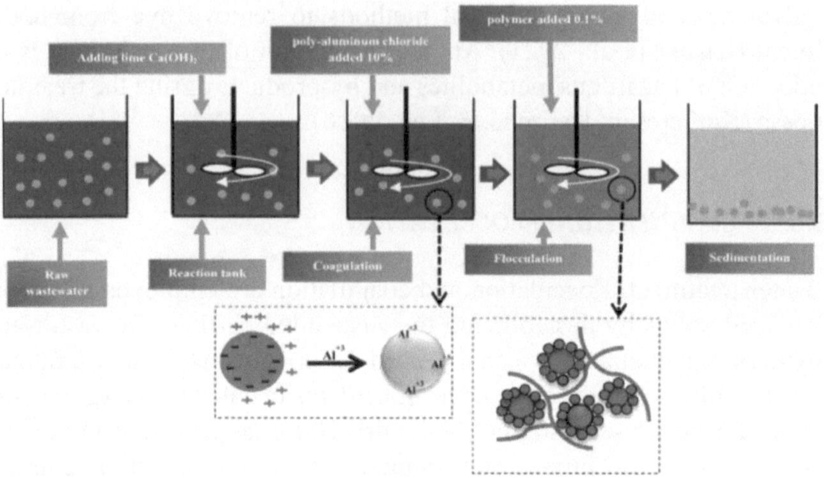

FIGURE 3.7 Schematic diagram for coagulation–flocculation process.

Source: Reprinted with permission from Ebrahimi et al., 2018. https://creativecommons.org/licenses/by/4.0/

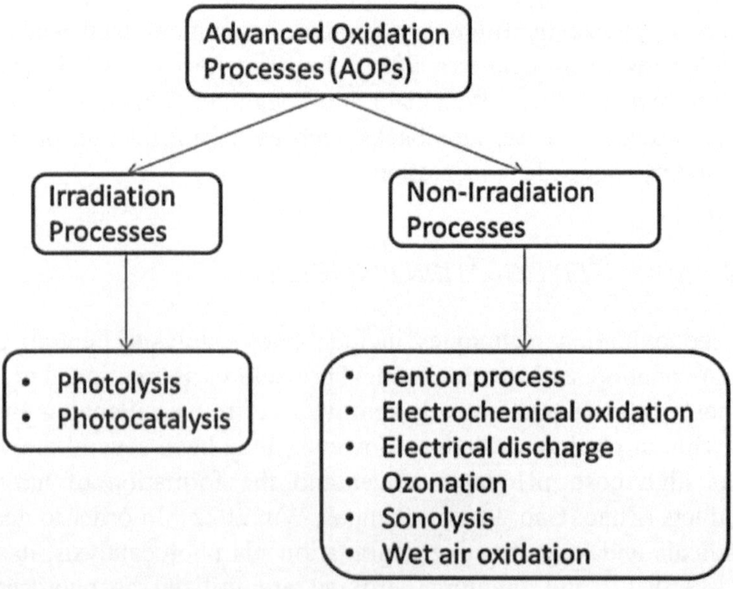

FIGURE 3.8 Different type of advanced oxidation processes.

Source: Reprinted from: Hisaindee, Meetani, & Rauf – Copyright (2013), with permission from Elsevier.

The most popular advanced oxidation techniques for breaking down organic contaminants in wastewater are the Fenton and photo-Fenton processes (Xiang, Ren, Yang, Xu, Zhang, & Wang, 2020). These methods rely on the presence of a combination of soluble iron (II) salt and Fenton's reagent (H_2O_2). The process consists of three steps: initiation, propagation, and termination (Li, Katz, & Hu, 2019). For treating wastewater containing dyes, a novel enhanced oxidation method called electrochemical technology has recently been developed, in which photo-electrocatalysis generates $OH^{\bullet}$ radicals at the anode surface (Mojiri, Ohashi, Ozaki, & Kindaichi, 2018).

3.3.2.3 ELECTROCHEMICAL APPROACHES

One of the promising solutions for removing organic and inorganic contaminants from wastewater is electrochemical water treatment. The majority of electrochemical water treatment technologies use electricity as the primary reactant to carry out the treatment process. There are different electrochemical water treatment technologies, such as electrocoagulation, electroflotation, electrodialysis, electrochemical oxidation, and reduction.

The electrocoagulation method utilizes two metallic electrodes for both *in-situ* coagulant particle production and direct power delivery, as shown in Figure 3.9. The cathode produces hydrogen gas, and the iron metal anode acts as a catalyst and coagulant. This process stands out for producing little sludge, being inexpensive, simple to use, and not requiring any chemicals (Hamad et al., 2018).

Another type of electrochemical technology includes direct and indirect processes, such as anodic oxidation. During the direct oxidation process, organic substances adsorbed at the surface of the anode are destroyed by the anodic e^- transfer mechanism. In the indirect oxidation process, electrochemistry is used to produce strong oxidants like O_3 and H_2O_2, resulting in the effective removal of colors and other organic contaminants, making this approach stand out (Montañés et al., 2020).

3.3.3 BIOLOGICAL TECHNIQUES

Biological approaches are more favorable than physical and chemical treatments because they are more practical, use fewer chemicals, are cheaper, consume less energy, are environmentally safe, and produce non-toxic byproducts. The advantage of biological techniques is the utilization of microorganisms with high biodegradability in wastewater treatment (Liu, You, Sun, Wang, Zhu, & Liu, 2018). Bacteria, algae, yeast, fungus, and enzyme-based systems are feasible biological options for converting dye molecules into non-toxic compounds in wastewater treatment. The different groups present on the cell wall of microbes, such as carboxyl, amino, and hydroxyl, are related to the mechanism for dye degradation (Jafari, Soudi, & Kasra-Kermanshahi, 2014).

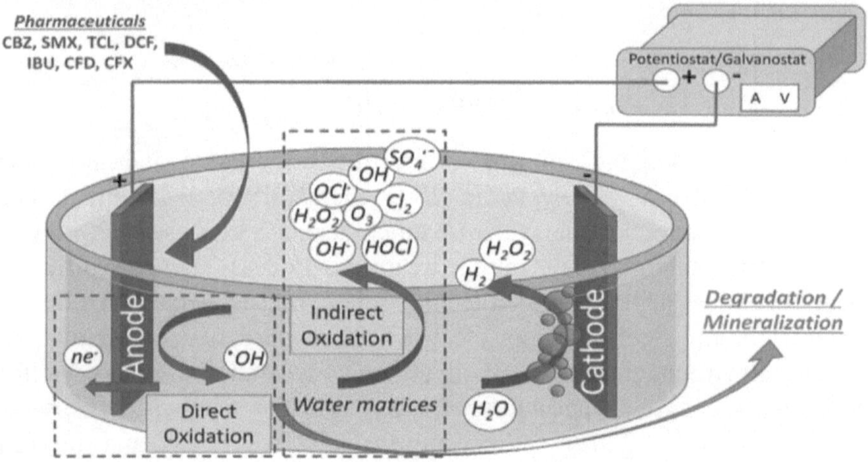

FIGURE 3.9 Process involved in the electrochemical oxidation process.

Source: Reprinted with permission from Bampos et al., 2021. https://creativecommons.org/licenses/by/4.0/

3.3.3.1 ENZYME-BASED DEGRADATION

Several investigations have been conducted to study the enzymatic activity involved in the breakdown of azo dyes. Among the available enzymes, peroxidase enzymes are widely used due to their high temperature stability and ability to degrade dyes over a broader pH range. Entrapping enzymes within a suitable matrix enhances the efficacy of dye treatment. The activity of the enzyme used varies depending on its class, source, and substrate. Furthermore, immobilized enzymes can be reused multiple times, and the decolorization effectiveness of immobilized enzymes is greater than that of soluble enzymes (Britos, Gianolini, Portillo, & Trelles, 2018). During the dye degradation process, enzymes such as manganese peroxidase and azo-reductase produce intermediates, such as phenols, which are more hazardous than the original dye molecules. In such cases, when two or more oxidoreductase enzymes combine, they can be utilized to transform harmful intermediates into insoluble end products.

3.3.3.2 MICROBIAL-ASSISTED DEGRADATION

The flexibility and aptitude of bacteria for the degradation of dye under the current environmental conditions affect the efficacy of dye degradation. The primary benefits of utilizing bacteria in degradation are their simplicity of culture and high rate of growth in comparison to other microbes (Kamali & Khodaparast, 2015). Bacteria show up to 100% efficacy in the biodegradation of dye. One of the researchers explored the decolorization of Indanthrene Blue RS using a bacterial consortium BP made up of Proteus mirabilis PMS, Bacillus flexus TS8, and Pseudomonas aeruginosa NCH. The BP consortium outperformed the individual strains, with an average decolorization rate (Mohanty & Kumar, 2021).

3.4 CHITOSAN (CH) AND CHITOSAN ANISALDEHYDE ENCAPSULATED V_2O_5 NANOPARTICLES FOR THE DEGRADATION OF METHYLENE BLUE FROM WASTEWATER

3.4.1 CHEMICALS

Chitosan (deacetylation $\geq$ 75%) was purchased from CDH Chemicals, Gallic acid (95%), Vanadyl sulfate (97%), Acetic acid, Sodium Borohydride, and p-Anisaldehyde were bought from SRL. All of the compounds were used as supplied.

3.4.2 SYNTHESIS

3.4.2.1 PREPARATION OF CHITOSAN-ANISALDEHYDE DERIVATIVE (CHA)

A 50 ml solution of CH (2 mg/L) in 1% acetic acid was prepared. To this, 5 ml of p-Anisaldehyde was added. The mixture was refluxed for 1 hour before the addition of $NaBH_4$. The pH of the solution was maintained at 8–9 using 0.1 M NaOH and then refluxed for 6 hours at 60°C. To remove the unreacted p-Anisaldehyde, the obtained precipitate was washed repeatedly with a mixture of ethanol and water. The final product, CHA, was dried under vacuum at 40°C (Chugh, Singh, Poddar, Thakur, Pani, & Jain, 2020).

3.4.2.2 FABRICATION OF ENCAPSULATED NANOPARTICLES

In a 250 ml beaker, 10 ml of CH or CHA was added to a 70 ml solution of 0.1 M vanadyl sulfate hydrate. The mixture was continuously stirred for 30 minutes at 80°C using a magnetic stirrer. After that, the colloidal solution was mixed for another hour while maintaining the pH at 8–9 using NaOH solution. Then, 10 ml of 0.4 M gallic acid solution was added and swirled for another 30 minutes. The resulting solution was cooled, and the obtained precipitate was washed with deionized water before being dried at 50°C for 24 hours (Bharathi, Ranjithkumar, Chandarshekar, & Bhuvaneshwari, 2019).

3.4.3 INSTRUMENTATION

FT-IR spectra of CH and CHA were obtained using a Perkin Elmer FT-IR spectrometer with a KBr pellet, in the scan range of 4,000–400 cm^{-1}. Further characterization was performed on a JEOL 400 MHz NMR spectrometer using ^1H NMR with D_2O/HCl as a mixed solvent.XRD analysis of the fabricated NPs was conducted on a Rigaku SmartLab 8A Series instrument, using Cu as an X-ray source in the scan range of 5–70°. The distinctive constituents of the NPs were studied using XPS (Thermo Fisher Scientific ESCALAB Xi+), with wide scan spectra ranging from 0 eV to 800 eV for binding energy. FE-SEM analysis (Carl Zeiss model Supra 55 Germany) was used to analyze particle size and surface morphology of the NPs. Prior to analysis, the samples were gold coated at a scanning voltage of 20 kV. UV-Visible spectroscopy (Esico international 3,577 twin beam spectrometer) was employed to determine the optical band gap for all of the NPs.

3.4.4 PHOTOCATALYTIC DEGRADATION OF METHYLENE BLUE

The photocatalytic activities of the synthesized NPs were assessed in an open environment by decolorization of methylene blue. V_2O_5 NPs were dispersed in an MB solution (50 mg/L). The maximum optical absorption of MB was measured at 664 nm using a UV-visible spectrometer (Esico international 3,577 double beam spectrometer), and the percentage of dye removal was calculated using Eqn. (1):

$$\text{Percent Dye Degraded} = \frac{C_0 - C}{C_0} \times 100 \qquad (1)$$

where; C_0 is the initial concentration of dye; and C is the concentration after time "t."

3.5 RESULTS AND DISCUSSION

3.5.1 CHARACTERIZATION OF CHITOSAN-ANISALDEHYDE

The FTIR spectra of CH and CHA are represented in Figure 3.10, with the major peaks mentioned in Table 3.1. Significant alterations in the FTIR spectra between the 1,250 cm^{-1}–1,650 cm^{-1} range are attributed to the formation of secondary amine of chitosan-anisaldehyde. The peak at 1,000 cm^{-1} indicates the presence of the aromatic ring of anisaldehyde, specifically due to the C=C bond bending vibration. The appearance of a peak at 1,630 cm^{-1} is attributed to the axial deformation of the N-H bond and the vibration of the new C-N bond, indicating the substitution of CH. The insertion of anisaldehyde onto CH increases the intensity of the band at 2,880 cm^{-1}. New peaks due to the ether group (Sajid et al., 2020) are observed at 1,305 cm^{-1} and 1,516 cm^{-1}.

In the ^{1}H-NMR spectra shown in Figure 3.11, the peaks corresponding to CHA are mentioned in Table 3.2. The peaks at 3.279 ppm and 3.015 ppm correspond to H_{11} and H_8, respectively, while the peaks in the range of 6–8 ppm correspond to the protons of benzene. The peak at 6.51 ppm indicates the presence of the C-N bond (Lamoureux, Singh, Jovic, Kuyyalil, Su, & Smith, 2016).

TABLE 3.1 FTIR Peaks for CH and CHA

Sample Code	Band (cm^{-1})	Functional Group and Type of Vibration
CH	3,922 cm^{-1}–3,000 cm^{-1}	OH stretching
	2,886 cm^{-1}	CH^- stretching
	1,661 cm^{-1}	CH_2 bending
	1,373 cm^{-1}	N–H bending
	1,152 cm^{-1}	C–O–C stretching
	1,059 cm^{-1}	C–O stretching
	986 cm^{-1}	NH_2 stretching
	822 cm^{-1}	C–N stretching
CHA	3,000 cm^{-1}–3,812 cm^{-1}	OH stretching
	2,873 cm^{-1}	CH^- stretching
	1,638 cm^{-1}	C=N stretching
	1,574 cm^{-1}	C–C stretching
	1,516 cm^{-1}	O–CH_3 bending
	1,463 cm^{-1}	N–H bending
	1,310 cm^{-1}	C–N stretching
	1,305 cm^{-1}	O–CH_3 stretching
	1,150 cm^{-1}	C–O–C stretching
	1,071 cm^{-1}	C–O stretching
	1,015 cm^{-1}	C=C bending
	897 cm^{-1}	C–N bending

3.5.2 CHARACTERIZATION OF ENCAPSULATED NANOPARTICLES

The characteristic V_2O_5 peaks were observed in XRD, confirming the fabrication of NPs. These peaks correspond to the orthorhombic crystalline phase of V_2O_5 NPs (ICDD-86-2248), as shown in Figure 3.12. Gallic acid (Kotov, 2017) exhibits characteristic peaks at 10.45°, 11.47°, and 23.77°, while CH shows a broad peak at $2\theta = 20°$. Additional CH peaks were found in CHA1, implying the presence of CH interacting with V_2O_5, indicating the involvement of an organic component in NP growth. The oxidation state of vanadium in NPs can be determined by analyzing the binding energy of the V2p level, specifically V 2p3/2 using XPS. The value was found to be 517.3 eV, confirming V^{+5} and validating the synthesis of V_2O_5. The main peak in V_2O_5 is centered at 530 eV (Figure 3.13), confirming

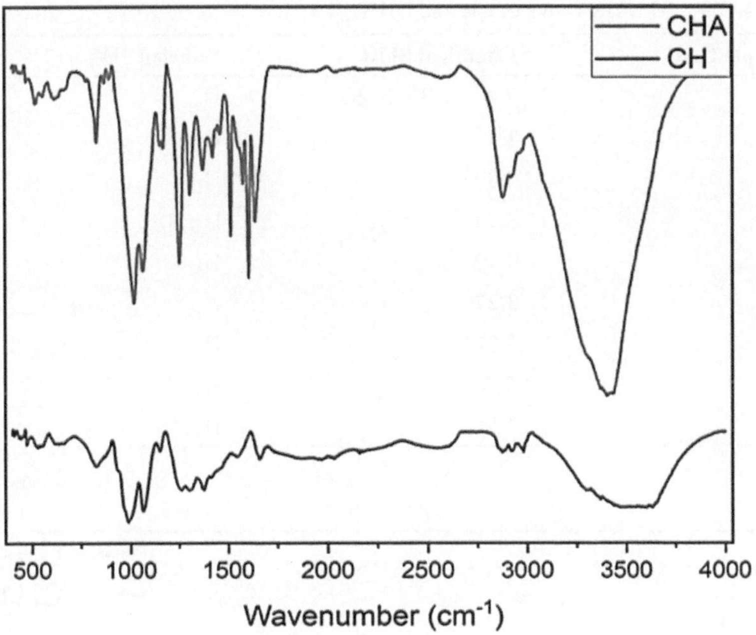

FIGURE 3.10 FTIR of CH and CHA.

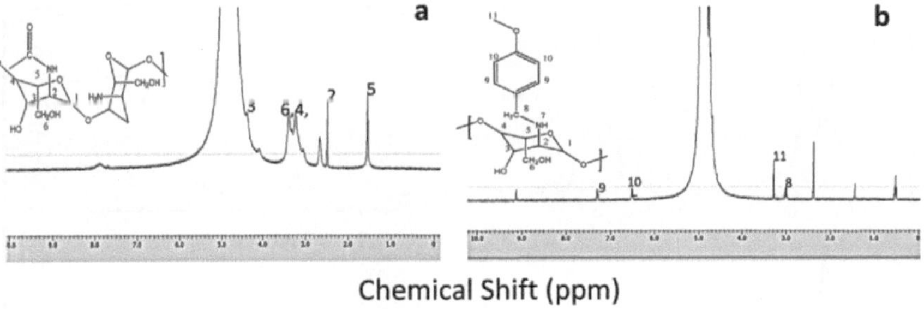

FIGURE 3.11 ^{1}H NMR of CH and CHA.

the presence of O^{2-} species on the NP surface. The signal at 532.1 eV is attributed to the C–O(H) bond, resulting from the binding of CH and CHA on the surface of the V_2O_5 NPs. Peaks at 284.8 eV are due to sp^3 hybridized carbon, indicating aromatic bonds in gallic acid. The peak at 400.7 eV in the N1s XPS spectrum is attributed to the C-N-C bonds (Jung & Oh, 2008).

TABLE 3.2 ^{1}H NMR Peaks of CH and CHA

Sample Code	Chemical Shift	Labeled "H"
CH	2.482, 2.39, 2.663	H_2
	1.5	H_5
	3.395	H_1
	3.24	H_4, H_6
	4.39	H_3
CHA	3.279	H_{11}
	3.015	H_8
	6–8	H_9, H_{10}
	6.51	H_7

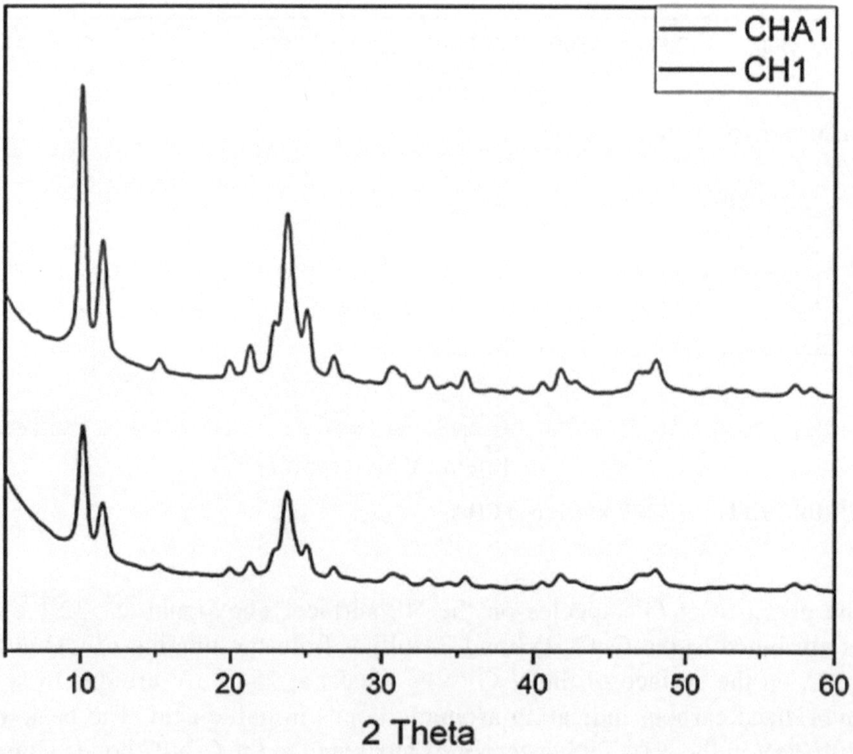

FIGURE 3.12 XRD pattern of CH1 and CHA1.

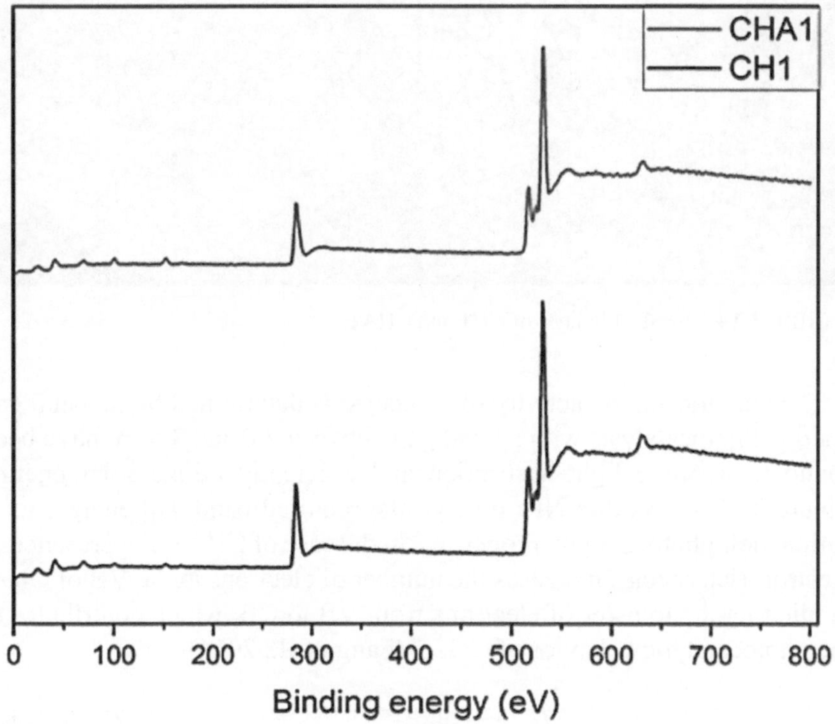

FIGURE 3.13 XPS pattern of CH1 and CHA1.

FESEM studies revealed that different CH derivatives resulted in a variety of morphologies, as shown in Figure 3.14. CH1 nanospheres with an average diameter of 26.0 ± 3.54 nm exhibited a nanosphere morphology. CHA1 possessed a rose-like structure with an average diameter of 971 ± 94.20 nm, with nanosheet thickness and width of 29.3 ± 3.73 nm and 199.2 ± 24.74 nm, respectively. In the case of CHA1, the $–OCH_3$ group in CHA promoted the creation of a rose-like structure due to 1–4 interaction, attributed to the electron-donating behavior of $–OCH_3$. These nanosheets formed a 3-D architecture due to the aggregation of NPs at the crystal interface, resulting in nanoplates that self-assemble into a hierarchical structure (Xu et al., 2020). These morphologies have a direct impact on photocatalytic performance due to surface features such as porosity and the presence of different functional groups. The inclusion of methoxy groups in CHA1, as well as its rosette structure, leads to the improvement of photocatalytic activity.

FIGURE 3.14 FESEM images of CH1 and CHA1.

The photocatalytic activity of a catalyst is determined by its band gap energy. Photocatalysts with a band gap between 2.0 and 3.5 eV have been found to improve light absorption and efficiently utilize solar energy. Figure 3.15 shows that NPs possess the required band gap energy, indicating their photocatalytic properties. In the case of CHA1, the presence of electron-rich species increases the number of electrons in the VB of CHA, facilitating the transfer of electrons from VB to CB, which contributes to the photocatalytic behavior of CHA1 (Fang et al., 2014).

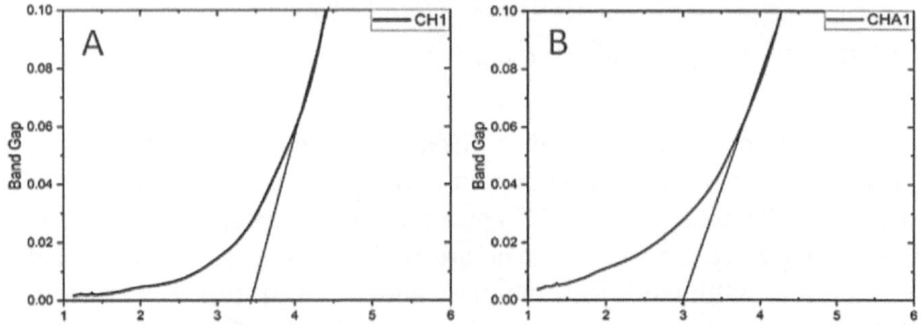

FIGURE 3.15 Optical band gap of CH1 and CHA1.

3.5.3 PHOTOCATALYTIC DEGRADATION OF METHYLENE BLUE

The photocatalytic activity of capped V_2O_5 nanoparticles (NPs) was investigated by degrading MB in an open environment. The highest absorbance at 664 nm was used to measure the degradation of MB. Prior to commencing the photocatalytic tests, adsorption/desorption equilibrium was achieved. It was found that it takes a maximum of 30 minutes to reach equilibrium. The degradation of MB using CH1 and CHA1 was 86.03% and 92.18% of the initial concentration, respectively, as shown in Figure 3.16.

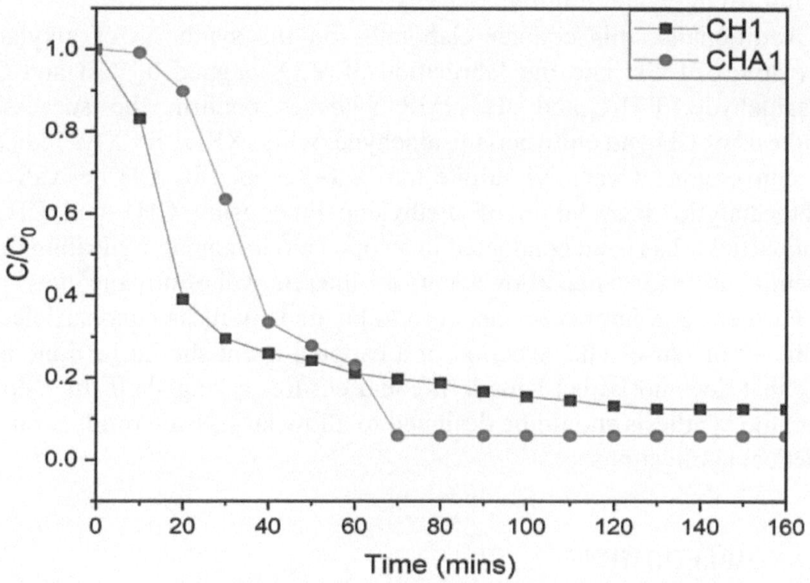

FIGURE 3.16 Graph representing photocatalytic degradation percent of MB.

CHA1 exhibited the highest degradation of 92%, which can be attributed to the morphology of the NPs, specifically the presence of rosette and wedge-shaped holes with improved crystallinity. This degradation occurs due to the reformation of the band gap caused by the addition of CH and CHA, which enhances charge separation and promotes the interaction between NPs and the irradiated light. Furthermore, active species such as free electrons and holes act as oxidant species in the breakdown of MB (Ma, Yang, Zhou, Xiao, Kennedy, & Nie, 2008).

3.6 CONCLUSIONS AND FUTURE PERSPECTIVES

This chapter discusses the harmful effects of textile dyes on living beings, including human beings, aquatic life, and plants. It also explains various physical, chemical, and biological methods that have been adopted for the removal of organic dyes. These techniques are based on different mechanisms, such as electrochemical processes, oxidation processes, membrane filtration, adsorption-based processes, and enzyme-based processes. All these methods are efficient, cost-effective, and promising for improving water quality.

Additionally, this chapter elaborates on the synthesis of alkylated derivatives of CH and the fabrication of V_2O_5 capped by CH and CH anisaldehyde. FTIR and ^1H NMR analyses confirm the successful synthesis of CH and chitosan-anisaldehyde, while XRD and XPS indicate the conversion of vanadyl sulfate into V_2O_5 capped by CH derivatives. Photocatalytic degradation of methylene blue using CH1 and CHA1 nanoparticles has been conducted in an open environment, indicating their potential as nano-remediation agents for the removal of organic dyes.

Furthermore, improvements need to be made in these nanoparticles to enable their reuse. The selection of a capping agent should be done in a way that does not affect aquatic life and ensures a long shelf life. Moreover, the synthesis should be designed to allow large-scale production for practical applications.

ACKNOWLEDGMENT

The authors express their gratitude to the Netaji University of Technology for providing the laboratory facilities and financial aid.

KEYWORDS

- **chitosan**
- **methylene blue**
- **nanoparticles**
- **organic dyes**
- **photocatalytic degradation**
- **vanadium pentoxide**

REFERENCES

Abu-Nada, A., Abdala, A., & McKay, G., (2021). Removal of phenols and dyes from aqueous solutions using graphene and graphene composite adsorption: A review. *J. Environ. Chem. Eng., 9*(5), 105858. doi: 10.1016/j.jece.2021.105858.

Affat, S. S., (2021). Classifications, advantages, disadvantages, toxicity effects of natural and synthetic dyes: A review. *UTJsci., 8*(1), 130–135. https://utjsci.utq.edu.iq/index.php/main (accessed on 24 February 2024).

Ahmad, A., Mohd-Setapar, S. H., Chuong, C. S., et al., (2015). Recent advances in new generation dye removal technologies: Novel search for approaches to reprocess wastewater. *RSC Adv., 5*(39), 30801–30818. doi: 10.1039/c4ra16959j.

Akpomie, K. G., & Conradie, J., (2020). Advances in application of cotton-based adsorbents for heavy metals trapping, surface modifications and future perspectives. *Ecotoxicol. Environ. Saf., 201*, 110825. doi: 10.1016/j.ecoenv.2020.110825.

Al-Mutairi, N. Z., (2006). Coagulant toxicity and effectiveness in a slaughterhouse wastewater treatment plant. *Ecotoxicol. Environ. Saf., 65*(1), 74–83. doi: 10.1016/j.ecoenv.2005.05.013.

Al-Tohamy, R., Ali, S. S., Li, F., et al., (2022). A critical review on the treatment of dye-containing wastewater: Ecotoxicological and health concerns of textile dyes and possible remediation approaches for environmental safety. *Ecotoxicol Environ. Saf., 231*, 113160. doi: 10.1016/j.ecoenv.2021.113160.

Amer, O. A., Ali, S. S., Azab, M., El-Shouny, W. A., Sun, J., & Mahmoud, Y. A. G., (2022). Exploring new marine bacterial species, *Alcaligenes faecalis* Alca F2018 valued for bioconversion of shrimp chitin to chitosan for concomitant biotechnological applications. *Int. J. Biol. Macromol., 196*, 35–45. doi: 10.1016/j.ijbiomac.2021.12.033.

Balakrishnan, V. K., Shirin, S., Aman, A. M., De Solla, S. R., Mathieu-Denoncourt, J., & Langlois, V. S., (2016). Genotoxic and carcinogenic products arising from reductive transformations of the azo dye, disperse yellow 7. *Chemosphere., 146*, 206–215. doi: 10.1016/j.chemosphere.2015.11.119.

Bashir, I., Lone, F. A., Bhat, R. A., Mir, S. A., Dar, Z. A., & Dar, S. A., (2020). Concerns and threats of contamination on aquatic ecosystems. *Bioremediation Biotechnol. Sustain. Approaches to Pollut. Degrad.* Published online, 1–26. doi: 10.1007/978-3-030-35691-0_1.

Bayramoglu, G., Kunduzcu, G., & Arica, M. Y., (2020). Preparation and characterization of strong cation exchange terpolymer resin as effective adsorbent for removal of disperse dyes. *Polym. Eng. Sci., 60*(1), 192–201. doi: 10.1002/pen.25272.

Behera, M., Nayak, J., Banerjee, S., Chakrabortty, S., & Tripathy, S. K., (2021). A review on the treatment of textile industry waste effluents towards the development of efficient mitigation strategy: An integrated system design approach. *J. Environ. Chem. Eng., 9*(4), 105277. doi: 10.1016/j.jece.2021.105277.

Berradi, M., Hsissou, R., Khudhair, M., et al., (2019). Textile finishing dyes and their impact on aquatic environs. *Heliyon., 5*(11). doi: 10.1016/j.heliyon.2019.e02711.

Bharathi, D., Ranjithkumar, R., Chandarshekar, B., & Bhuvaneshwari, V., (2019). Preparation of chitosan coated zinc oxide nanocomposite for enhanced antibacterial and photocatalytic activity: As a bionanocomposite. *Int. J. Biol. Macromol., 129*, 989–996. doi: 10.1016/j.ijbiomac.2019.02.061.

Britos, C. N., Gianolini, J. E., Portillo, H., & Trelles, J. A., (2018). Biodegradation of industrial dyes by a solvent, metal, and surfactant-stable extracellular bacterial laccase. *Biocatal. Agric. Biotechnol., 14*, 221–227. doi: 10.1016/j.bcab.2018.03.015.

Burakov, A. E., Galunin, E. V., Burakova, I. V., et al., (2018). Adsorption of heavy metals on conventional and nanostructured materials for wastewater treatment purposes: A review. *Ecotoxicol. Environ. Saf., 148*, 702–712. doi: 10.1016/j.ecoenv.2017.11.034.

Cao, G., Wang, R., Ju, Y., et al., (2021). Synchronous removal of emulsions and soluble organic contaminants via a microalgae-based membrane system: Performance and mechanisms. *Water Res., 206*, 117741. doi: 10.1016/j.watres.2021.117741.

Carney, A. B., Cartine, J., Jönander, C., et al., (2021). Assessing the effects of textile leachates in fish using multiple testing methods: From gene expression to behavior. *Ecotoxicol. Environ. Saf., 207*. doi: 10.1016/j.ecoenv.2020.111523.

Chan, S. H. S., Wu, T. Y., Juan, J. C., & Teh, C. Y., (2011). Recent developments of metal oxide semiconductors as photocatalysts in advanced oxidation processes (AOPs) for treatment of dye waste-water. *J. Chem. Technol. Biotechnol., 86*(9), 1130–1158. doi: 10.1002/jctb.2636.

Chugh, B., Singh, A. K., Poddar, D., Thakur, S., Pani, B., & Jain, P., (2020). Relation of degree of substitution and metal protecting ability of cinnamaldehyde modified chitosan. *Carbohydr. Polym., 234*, 115945. doi: 10.1016/j.carbpol.2020.115945.

Copaciu, F., Opriş, O., Coman, V., Ristoiu, D., Niinemets, Ü., & Copolovici, L., (2013). Diffuse water pollution by anthraquinone and azo dyes in environment importantly alters foliage volatiles, carotenoids, and physiology in wheat (*Triticum aestivum*). *Water Air Soil Pollut., 224*(3). doi: 10.1007/s11270-013-1478-4.

Danish, K. M., Abdulateif, H., Ismail, I. M., Sabir, S., & Zain, K. M., (2015). Bioelectricity generation and bioremediation of an azo-dye in a microbial fuel cell coupled activated sludge process. *PLoS One, 10*(10). doi: 10.1371/journal.pone.0138448.

Dasgupta, J., Mondal, D., Chakraborty, S., Sikder, J., Curcio, S., & Arafat, H. A., (2015). Nanofiltration based water reclamation from tannery effluent following coagulation pretreatment. *Ecotoxicol. Environ. Saf., 121*, 22–30. doi: 10.1016/j.ecoenv.2015.07.006.

Dasgupta, J., Singh, M., Sikder, J., Padarthi, V., Chakraborty, S., & Curcio, S., (2015). Response surface-optimized removal of reactive red 120 dye from its aqueous solutions using polyethyleneimine enhanced ultrafiltration. *Ecotoxicol. Environ. Saf., 121*, 271–278. doi: 10.1016/j.ecoenv.2014.12.041.

Fang, Y., Li, Z., Yang, B., et al., (2014). Effect of dye structure on optical properties and photocatalytic behaviors of squaraine-sensitized TiO_2 nanocomposites. *J Phys Chem C., 118*(29), 16113–16125. doi: 10.1021/jp502208y.

Ferraz, E. R. A., Grando, M. D., & Oliveira, D. P., (2011). The azo dye disperse orange 1 induces DNA damage and cytotoxic effects but does not cause ecotoxic effects in *Daphnia similis* and *Vibrio fischeri*. *J. Hazard Mater., 192*(2), 628–633. doi: 10.1016/j.jhazmat.2011.05.063.

Field, C. B., Campbell, J. E., & Lobell, D. B., (2008). Biomass energy: The scale of the potential resource. *Trends Ecol. Evol., 23*(2), 65–72. doi: 10.1016/j.tree.2007.12.001.

Gelagutashvili, E., (2013). Comparative study on heavy metals biosorption by different types of bacteria. *Open J. Met., 3*(2), 62–67. doi: 10.4236/ojmetal.2013.32a1008.

Ghaedi, M., Hajjati, S., Mahmudi, Z., et al., (2015). Modeling of competitive ultrasonic assisted removal of the dyes – methylene blue and safranin-O using Fe_3O_4 nanoparticles. *Chem Eng J., 268*, 28–37. doi: 10.1016/j.cej.2014.12.090.

Hamad, H., Bassyouni, D., El-Ashtoukhy, E. S., Amin, N., & Abd El-Latif, M., (2018). Electrocatalytic degradation and minimization of specific energy consumption of synthetic azo dye from wastewater by anodic oxidation process with an emphasis on enhancing economic efficiency and reaction mechanism. *Ecotoxicol. Environ. Saf., 148*, 501–512. doi: 10.1016/j.ecoenv.2017.10.061.

Hassanshahi, N., & Karimi-Jashni, A., (2018). Comparison of photo-Fenton, O_3 /H_2O_2 /UV and photocatalytic processes for the treatment of gray water. *Ecotoxicol. Environ. Saf., 161*, 683–690. doi: 10.1016/j.ecoenv.2018.06.039.

Hisaindee, S., Meetani, M. A., & Rauf, M. A., (2013). Application of LC-MS to the analysis of advanced oxidation process (AOP) degradation of dye products and reaction mechanisms. *TrAC Trends in Analytical Chemistry, 49*, 31–44.

Hossain, M. S., Das, S. C., Islam, J. M. M., Al Mamun, M. A., & Khan, M. A., (2018). Reuse of textile mill ETP sludge in environmental friendly bricks – effect of gamma radiation. *Radiat. Phys. Chem., 151*, 77–83. doi: 10.1016/j.radphyschem.2018.05.020.

Islam, T., Repon, M. R., Islam, T., Sarwar, Z., & Rahman, M. M., (2022). *Impact of Textile Dyes on Health and Ecosystem: A Review of Structure, Causes, and Potential Solutions*. Springer Berlin Heidelberg. doi: 10.1007/s11356-022-24398-3.

Jafari, N., Soudi, M. R., & Kasra-Kermanshahi, R., (2014). Biodegradation perspectives of azo dyes by yeasts. *Microbiol. (Russian Fed.), 83*(5), 484–497. doi: 10.1134/S0026261714050130.

Jung, S., & Oh, M., (2008). Monitoring shape transformation from nanowires to nanocubes and size-controlled formation of coordination polymer particles. *Angew Chemie – Int. Ed., 47*(11), 2049–2051. doi: 10.1002/anie.200704209.

Kamali, M., & Khodaparast, Z., (2015). Review on recent developments on pulp and paper mill wastewater treatment. *Ecotoxicol. Environ. Saf., 114*, 326–342. doi: 10.1016/j.ecoenv.2014.05.005.

Katheresan, V., Kansedo, J., & Lau, S. Y., (2018). Efficiency of various recent wastewater dye removal methods: A review. *J. Environ. Chem. Eng., 6*(4), 4676–4697. doi: 10.1016/j.jece.2018.06.060.

Kaushik, P., Garg, V. K., & Singh, B., (2005). Effect of textile effluents on growth performance of wheat cultivars. *Bioresour. Technol., 96*(10), 1189–1193. doi: 10.1016/j.biortech.2004.09.020.

Kishor, R., Purchase, D., Saratale, G. D., et al., (2021). Ecotoxicological and health concerns of persistent coloring pollutants of textile industry wastewater and treatment approaches for environmental safety. *J. Environ. Chem. Eng., 9*(2), 105012. doi: 10.1016/j.jece.2020.105012.

Korpi, A., Järnberg, J., & Pasanen, A. L., (2009). Microbial volatile organic compounds. *Crit. Rev. Toxicol., 39*(2), 139–193. doi: 10.1080/10408440802291497.

Kotov, N. A., (2017). Self-assembly of inorganic nanoparticles: Ab Ovo. *EPL (Europhysics Lett., 119*(6), 66008. doi: 10.1209/0295-5075/119/66008.

Lamoureux, B., Singh, V. R., Jovic, V., Kuyyalil, J., Su, T. Y., & Smith, K. E., (2016). Structural and electronic properties of thermally evaporated V_2O_5 epitaxial thin films. *Thin Solid Films., 615*, 409–414. doi: 10.1016/j.tsf.2016.07.062.

Lellis, B., Fávaro-Polonio, C. Z., Pamphile, J. A., & Polonio, J. C., (2019). Effects of textile dyes on health and the environment and bioremediation potential of living organisms. *Biotechnol. Res. Innov., 3*(2), 275–290. doi: 10.1016/j.biori.2019.09.001.

Leme, D. M., Primo, F. L., Gobo, G. G., Da Costa, C. R. V., Tedesco, A. C., De Oliveira, D. P., (2015). Genotoxicity assessment of reactive and disperse textile dyes using human dermal equivalent (3D cell culture system). *J. Toxicol. Environ. Heal – Part A Curr. Issues, 78*(7), 466–480. doi: 10.1080/15287394.2014.999296.

Li, F., Katz, L., & Hu, Z., (2019). Adsorption of major nitrogen-containing components in microalgal bio-oil by activated carbon: Equilibrium, kinetics, and ideal adsorbed solution theory (IAST) model. *ACS Sustain Chem Eng*. doi: 10.1021/acssuschemeng.9b03804.

Li, H., Zhang, R., Tang, L., Zhang, J., & Mao, Z., (2014). Evaluation of *Bacillus* sp. MZS10 for decolorizing azure B dye and its decolorization mechanism. *J. Environ. Sci. (China), 26*(5), 1125–1134. doi: 10.1016/S1001-0742(13)60540-9.

Li, J., Zheng, B. H., He, Y., et al., (2018). Antimony contamination, consequences, and removal techniques: A review. *Ecotoxicol. Environ. Saf., 156*, 125–134. doi: 10.1016/j.ecoenv.2018.03.024.

Liang, J., Ning, X., Kong, M., et al., (2017). Elimination and ecotoxicity evaluation of phthalic acid esters from textile-dyeing wastewater. *Environ Pollut., 231*, 115–122. doi: 10.1016/j.envpol.2017.08.006.

Liu, W., You, Y., Sun, D., Wang, S., Zhu, J., & Liu, C., (2018). Decolorization and detoxification of water-insoluble Sudan dye by *Shewanella putrefaciens* CN32 co-cultured with *Bacillus circulans* BWL1061. *Ecotoxicol Environ Saf., 166*(101), 11–17. doi: 10.1016/j.ecoenv.2018.09.055.

Ma, G., Yang, D., Zhou, Y., Xiao, M., Kennedy, J. F., & Nie, J., (2008). Preparation and characterization of water-soluble N-alkylated chitosan. *Carbohydr Polym., 74*(1), 121–126. doi: 10.1016/j.carbpol.2008.01.028.

Madan, S., Shaw, R., Tiwari, S., & Tiwari, S. K., (2019). Adsorption dynamics of Congo red dye removal using ZnO functionalized high silica zeolitic particles. *Appl. Surf. Sci., 487*, 907–917. doi: 10.1016/j.apsusc.2019.04.273.

Mani, S., & Bharagava, R. N., (2016). Exposure to Crystal Violet, Its Toxic, Genotoxic and Carcinogenic Effects on Environment and Its Degradation and Detoxification for Environmental Safety, 237. doi: 10.1007/978-3-319-23573-8_4.

Moawad, H., Abd El-Rahim, W. M., & Khalafallah, M., (2003). Evaluation of biotoxicity of textile dyes using two bioassays. *J. Basic Microbiol., 43*(3), 218–229. doi: 10.1002/jobm.200390025.

Mohanty, S. S., & Kumar, A., (2021). Enhanced degradation of anthraquinone dyes by microbial monoculture and developed consortium through the production of specific enzymes. *Sci. Rep., 11*(1), 1–15. doi: 10.1038/s41598-021-87227-6.

Mojiri, A., Ohashi, A., Ozaki, N., & Kindaichi, T., (2018). Pollutants removal from synthetic wastewater by the combined electrochemical, adsorption and sequencing batch reactor (SBR). *Ecotoxicol. Environ. Saf., 161*, 137–144. doi: 10.1016/j.ecoenv.2018.05.053.

Montañés, M. T., García-Gabaldón, M., Roca-Pérez, L., Giner-Sanz, J. J., Mora-Gómez, J., & Pérez-Herranz, V., (2020). Analysis of norfloxacin ecotoxicity and the relation with its degradation by means of electrochemical oxidation using different anodes. *Ecotoxicol. Environ. Saf., 188*, 109923. doi: 10.1016/j.ecoenv.2019.109923.

Mudhoo, A., Ramasamy, D. L., Bhatnagar, A., Usman, M., & Sillanpää, M., (2020). An analysis of the versatility and effectiveness of composts for sequestering heavy metal ions, dyes and xenobiotics from soils and aqueous milieus. *Ecotoxicol. Environ. Saf., 197*, 110587. doi: 10.1016/j.ecoenv.2020.110587.

Pei, S., Wang, Y., You, S., Li, Z., & Ren, N., (2022). Electrochemical removal of chlorophenol pollutants by reactive electrode membranes: Scale-up strategy for engineered applications. *Engineering., 9*, 77–84. doi: 10.1016/j.eng.2021.11.017.

Peng, H., & Guo, J., (2020). Removal of chromium from wastewater by membrane filtration, chemical precipitation, ion exchange, adsorption electrocoagulation, electrochemical reduction, electrodialysis, electrodeionization, photocatalysis and nanotechnology: A review. *Environ. Chem. Lett., 18*(6), 2055–2068. doi: 10.1007/s10311-020-01058-x.

Piątkowska, M., Jedziniak, P., Olejnik, M., Żmudzki, J., & Posyniak, A., (2018). Absence of evidence or evidence of absence? A transfer and depletion study of Sudan I in eggs. *Food Chem., 239*, 598–602. doi: 10.1016/j.foodchem.2017.06.156.

Rawat, D., Sharma, R. S., Karmakar, S., Arora, L. S., & Mishra, V., (2018). Ecotoxic potential of a presumably non-toxic azo dye. *Ecotoxicol. Environ. Saf., 148*, 528–537. doi: 10.1016/j.ecoenv.2017.10.049.

Ribeiro, A. R., & Umbuzeiro, G. De. A., (2014). Effects of a textile azo dye on mortality, regeneration, and reproductive performance of the planarian, *Girardia tigrina*. *Environ. Sci. Eur., 26*(1), 1–8. doi: 10.1186/s12302-014-0022-5.

Sajid, M. M., Shad, N. A., Javed, Y., et al., (2020). Preparation and characterization of vanadium pentoxide (V_2O_5) for photocatalytic degradation of monoazo and diazo dyes. *Surfaces and Interfaces., 19*, 100502. doi: 10.1016/j.surfin.2020.100502.

Samsami, S., Mohamadi, M., Sarrafzadeh, M. H., Rene, E. R., & Firoozbahr, M., (2020). Recent advances in the treatment of dye-containing wastewater from textile industries: Overview and perspectives. *Process Saf. Environ. Prot., 143*, 138–163. doi: 10.1016/j.psep.2020.05.034.

Sarvajith, M., Reddy, G. K. K., & Nancharaiah, Y. V., (2018). Textile dye biodecolourization and ammonium removal over nitrite in aerobic granular sludge sequencing batch reactors. *J Hazard Mater., 342*, 536–543. doi: 10.1016/j.jhazmat.2017.08.064.

Sivarajasekar, N., & Baskar, R., (2014). Adsorption of basic red 9 on activated waste *Gossypium hirsutum* seeds: Process modeling, analysis and optimization using statistical design. *J Ind Eng Chem., 20*(5), 2699–2709. doi: 10.1016/j.jiec.2013.10.058.

Siwińska-Stefańska, K., Kubiak, A., Piasecki, A., et al., (2018). TiO_2-ZnO binary oxide systems: Comprehensive characterization and tests of photocatalytic activity. *Materials (Basel), 11*(5), 1–19. doi: 10.3390/ma11050841.

Srivastava, K., Kumar, V., Alam, A., et al., (2020). Impact of the dams on water and plankton quality of the River Ganga (Gangotri to Vindhyachal). *World Water Policy, 6*(2), 299–313. doi: 10.1002/wwp2.12038.

Sunil, K., Sherugar, P., Rao, S., Lavanya, C., Balakrishna, G. R., Arthanareeswaran, G., & Padaki, M., (2021). Prolific approach for the removal of dyes by an effective interaction with polymer matrix using ultrafiltration membrane. *Journal of Environmental Chemical Engineering*, 9/6.

Suryavathi, V., Sharma, S., Sharma, S., et al., (2005). Acute toxicity of textile dye wastewaters (untreated and treated) of Sanganer on male reproductive systems of albino rats and mice. *Reprod. Toxicol., 19*(4), 547–556. doi: 10.1016/j.reprotox.2004.09.011.

Tkaczyk, A., Mitrowska, K., & Posyniak, A., (2020). Synthetic organic dyes as contaminants of the aquatic environment and their implications for ecosystems: A review. *Science of The Total Environment*, 717/137222.

Tounsadi, H., Metarfi, Y., Taleb, M., El Rhazi, K., & Rais, Z., (2020). Impact of chemical substances used in textile industry on the employee's health: Epidemiological study. *Ecotoxicol. Environ. Saf., 197*. doi: 10.1016/j.ecoenv.2020.110594.

Xiang, H., Ren, G., Yang, X., Xu, D., Zhang, Z., & Wang, X., (2020). A low-cost solvent-free method to synthesize α-Fe_2O_3 nanoparticles with applications to degrade methyl orange in photo-Fenton system. *Ecotoxicol Environ Saf., 200*, 110744. doi: 10.1016/j.ecoenv.2020.110744.

Xu, L., Zhang, J., Ding, J., et al., (2020). Pore structure and fractal characteristics of different shale lithofacies in the dalong formation in the western area of the lower Yangtze platform. *Minerals., 10*(1). doi: 10.3390/min10010072.

Zarei, M. T., Ebrahimi, A. A., Eslami, H., et al., (2018). Optimization and economic evaluation of modified coagulation–flocculation process for enhanced treatment of ceramic-tile industry wastewater. *AMB Express, 8*(1). doi: 10.1186/s13568-018-0702-4.

Zhao, H., Cao, P., Lu, L., Li, F., Pang, C. H., & Wu, T., (2022). Co-regulation of dispersion, exposure and defect sites on CeO_2 (111) surface for catalytic oxidation of Hg^0. *J Hazard Mater., 424*. doi: 10.1016/j.jhazmat.2021.126566.

CHAPTER 4

Pesticides Removal from Water with Semiconductor Materials Through Adsorption and Photocatalysis: A Sustainable Strategy

SOFÍA ESTRADA FLORES and ANTONIA MARTÍNEZ LUÉVANOS

Department of Advanced Ceramic Materials and Energy, Facultad de Ciencias Químicas, Universidad Autónoma de Coahuila, Saltillo, Coahuila, México

ABSTRACT

Nowadays, anthropogenic activities have caused a reduction in potable water sources due to contamination and overexploitation. Agriculture is one of the human activities that generates the most wastewater with high amounts of pollutants. Pesticides are the main type of pollutants found in this type of wastewater, and their removal is important due to the negative effects they can cause in humans and other living beings. Semiconductor materials, thanks to their electrical, optical, and chemical properties, can be used for the elimination of these organic compounds from aqueous media. Materials such as metal oxides, certain sulfurs, and their heterojunctions have been tested as adsorbents and photocatalysts for the degradation of pesticides with good efficiency. However, researchers are still working on improving the morphological and optical properties of these materials. In this chapter, we will provide an overview of different types of semiconductor materials and the properties that make them suitable for

Sustainable Environment and Health: Practical Strategies. Leonardo Sepúlveda Torre, Ann Rose Abraham, Neha Kanwar Rawat, & A. K. Haghi (Eds.)

the removal of pesticides in water through processes like adsorption and photocatalysis.

4.1 INTRODUCTION TO PESTICIDES IN WATER

Currently, the sources of drinking water have been diminished due to the increasing anthropogenic activities and population growth. As a result, access to clean water has become a problem in the last couple of decades. This issue has caught the attention of various organizations worldwide, including the United Nations (UN), which has set a specific goal in the sustainable development goals (SDGs) related to clean water and sanitation. One way to achieve this goal is by reducing the pollutants present in water (Sustainable Development Goals of the United Nations Development Program, 2015).

The scientific community has developed methods for removing water pollutants. However, there are some substances considered to be emerging and persistent in water sources. Pesticides are among these substances, as they are used to control pests and aid in plant growth (Bano, Kaushal, & Singh, 2021; Yeganeh, Charkhloo, Sobhi, Esrafili, & Gholami, 2022). Currently, the use of pesticides in agriculture is essential, and the presence of pesticides in wastewater and drinking water has become more common (El-Nahhal & El-Nahhal, 2021).

Despite a decrease in pesticide consumption in recent years, global consumption has actually increased over the last decade, as shown in Figure 4.1. According to the Food and Agricultural Organization of the United Nations, the countries with the highest consumption levels are the United States of America, Brazil, and China. Figure 4.2 displays the annual pesticide consumption by country (Data of the Food and Agriculture Organization of the United Nations, 2020).

According to their chemical composition, pesticides can be classified as carbamates, organophosphates, organochlorine, and pyrethroids (Ore, Adeola, Bayode, Adedipe, & Nomngongo, 2023). Figure 4.3 shows the structure of some commercial pesticides. In 2021, El-Nahhal reported an analysis of different papers on pesticide occurrence in water, with DDT (dichlorodiphenyltrichloroethane) being the most common. Highly toxic organophosphates were also found in drinking water (El-Nahhal & El-Nahhal, 2021). Despite the existence of processes for the removal of these substances, such as advanced oxidation and adsorption, there is a need to improve the materials used for this purpose. Properties like morphology, texture, and optical properties directly affect the removal percentage of pollutants from aqueous media (Estrada-Flores et al., 2020).

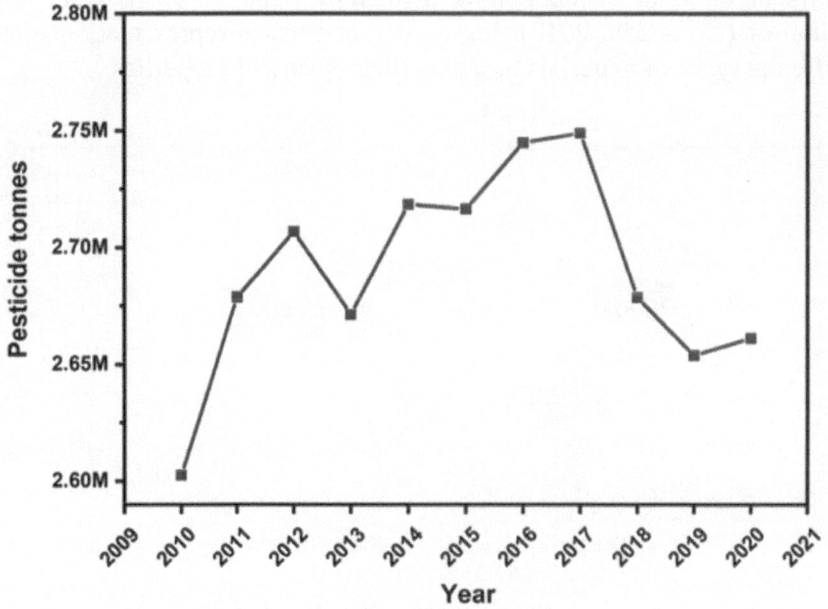

FIGURE 4.1 *Annual global consumption of pesticides from 2010–2020 according to the FAOSTAT (Food and Agricultural Organization Statistics).*

Due to their optical properties, semiconductor materials have been used for the removal of organic pollutants from water through techniques such as adsorption and photocatalysis (Wei, Xu, Gao, Liu, & Huang, 2012; Yadav, Kumar, Kumari, Mittal, & Sharma, 2019; Sharma, Hazra, & Basu, 2017; Ilina et al., 2017). In the following lines, the general properties of semiconductors and their use in these processes will be described.

4.2 INTRODUCTION TO SEMICONDUCTOR MATERIALS

A semiconductor is a material with electrical properties that lie between those of a conductor and an insulator. In a semiconductor, the valence band and the conduction band are separated by an energy state called the band gap (Eg), where electrons cannot be found. The band gap value in a semiconductor must be higher than 0 eV but lower than 4 eV; materials with a band gap higher than 4 eV are considered insulators. The low value of the band gap in semiconductors allows electrons in the valence band to transition to the conduction band when the material receives an energy stimulus (Coronado, 2013). Figure 4.4 presents a representation of the different types of materials based on their electrical properties.

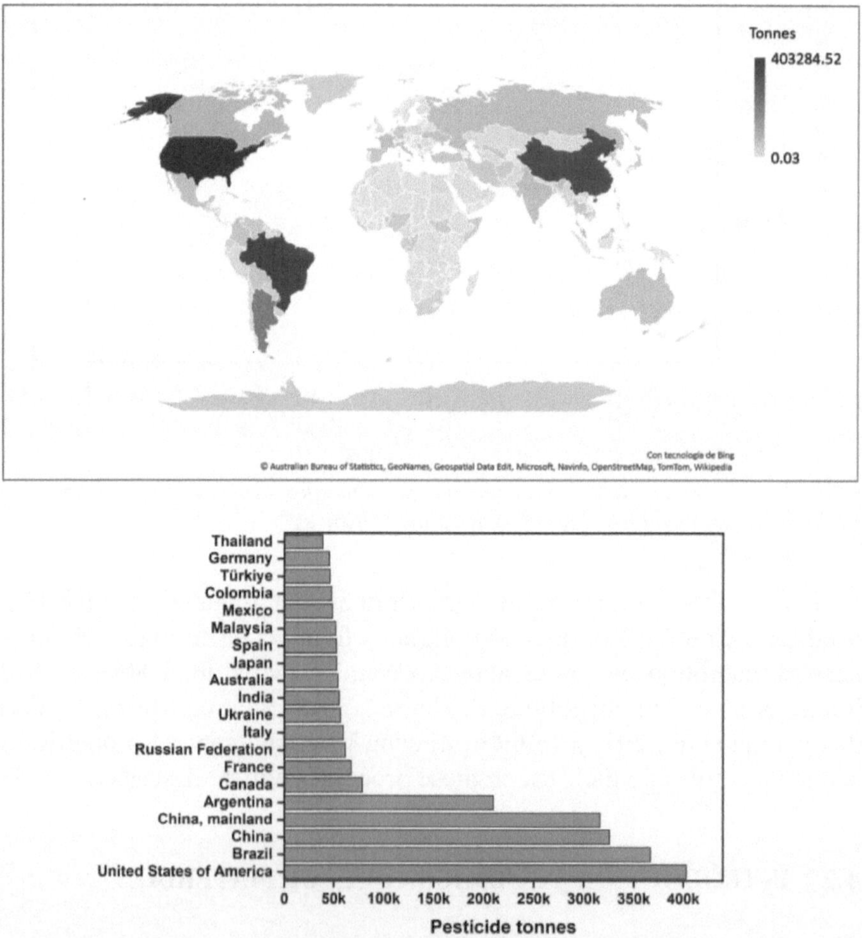

FIGURE 4.2 Worldwide use of pesticides and average annual consumption of pesticides by country according to the FAOSTAT from 2010–2020.

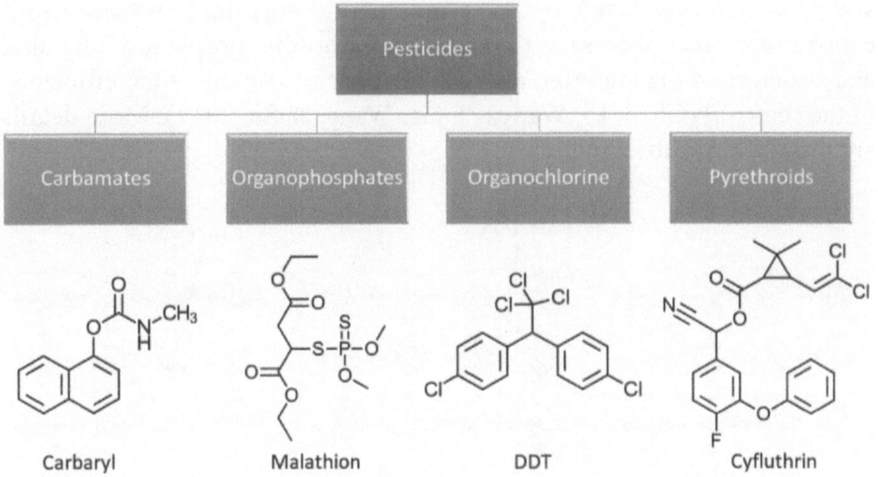

FIGURE 4.3 Different type of pesticides according to their chemical composition.

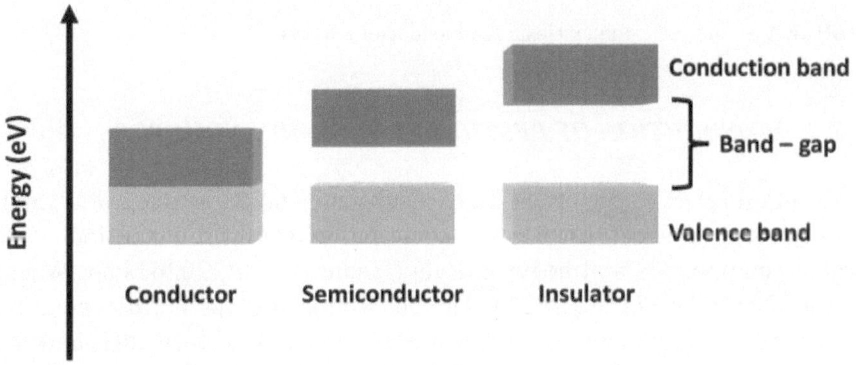

FIGURE 4.4 Types of materials according to their electric properties.

Semiconductors can be classified as intrinsic and extrinsic; the former are pure semiconductors, while the latter are semiconductors containing impurities, also known as doped semiconductors. These modified semiconductors can be n-type or p-type. As shown in Figure 4.5, n-type semiconductors are doped with elements that provide electrons (–), while p-type semiconductors are doped with elements that provide holes (+).

As mentioned above, the optical and electrical properties of semiconductor materials make them suitable for the removal of organic pollutants from water. The most common processes for this purpose are adsorption and photocatalysis. While optical properties are important for these applications, it is also necessary to pay attention to the morphology, texture, and structure of the material, as these factors can also affect the efficiency of the removal (Chen, Li, Wang, Zhang, Wang, & Su, 2012). More details are provided in subsections.

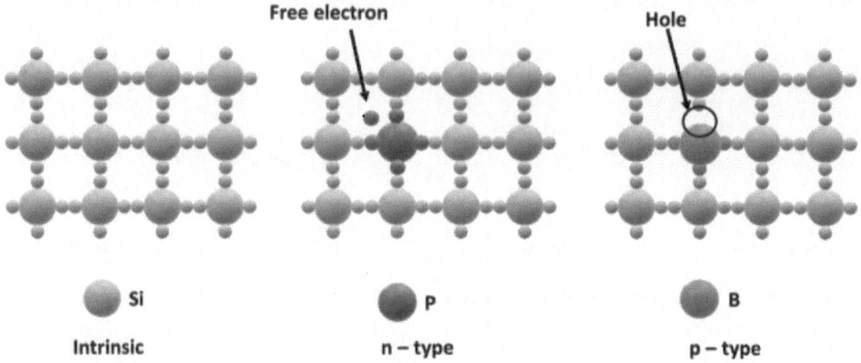

FIGURE 4.5 Intrinsic and extrinsic semiconductor materials.

4.2.1 ADSORPTION OF PESTICIDES BY SEMICONDUCTORS

Adsorption refers to the retention of a substance on the surface of a solid. This process has been used to eliminate pollutants from water, including organic compounds and heavy metals (Hitam & Jalil, 2020; Pan, Wang, & Liu, 2012; Tho et al., 2021). The adsorption process is influenced by factors such as temperature, concentration of the adsorbent, pH, and the physical and chemical properties of the adsorbent. Some of these properties are described in succeeding sections.

4.2.2 PROPERTIES OF A SEMICONDUCTOR THAT AFFECTS THE ADSORPTION PROCESS

4.2.2.1 TEXTURE AND MORPHOLOGY

The specific surface area and morphology of the material are highly related to the adsorption capacity. A semiconductor with a high specific surface area can adsorb a greater amount of pollutants compared to one with a low specific surface area. The presence of pores increases the specific surface area; however, the type of pores can also affect the adsorption process. Some substances are preferably adsorbed onto small pores such as micropores (with an average diameter lower than 2 nm) or mesopores (with an average diameter between 2 and 50 nm) rather than macropores (with an average diameter higher than 50 nm) (Gerasimova et al., 2016).

The size of the particles can also affect the adsorption capacity. Smaller particles commonly have larger surface areas, and an increase in specific surface area typically leads to higher adsorption. Therefore, the adsorption efficiency is inversely proportional to the particle size (Mohammed & Ali, 2023).

The shape of the particles is another important characteristic in adsorption. Recently, nanotube arrays have attracted attention due to their improved properties resulting from the formation of multiple vertical channels that increase the specific surface area (Zhou & Fang, 2015). The shape of a semiconductor can be controlled by the use of surfactants that act as templates through the formation of micelles. For example, different shapes of TiO_2, such as porous agglomerates, mono-layered, and bi-layered particles, have been obtained using ionic and non-ionic surfactants like SDS (sodium dodecyl sulfate), CTAB, and PEG (polyethylene glycol) (Estrada-Flores et al., 2020; Nagamine & Sasaoka, 2002; Li, He, Xiao, Liu, & Wang, 2009).

4.2.2.2 ISOELECTRIC POINT

Another property related to the adsorption capacity is the isoelectric point of the semiconductor. When a non-soluble solid material comes into contact with a liquid, it develops a surface charge that varies with the pH of the medium. The pH value at which this surface charge is zero is known as the isoelectric point, which is specific to each semiconductor. If the pH is higher than the isoelectric point, a negatively charged surface is formed, while a positively charged surface is formed if the pH is lower. As a result, a semiconductor can have different adsorption capacities for the same substance at different pH values.

In 2014, Brigante et al. reported the adsorption of paraquat (PQ^{2+}) with SiO_2. The results indicated that the highest adsorption efficiency was achieved at pH 9.5, which decreased at pH 7 and further at pH 4.4. The decrease at pH 4.4 was attributed to the isoelectric point of SiO_2, which is around 2. With an increase in pH, the amount of the cationic form of the pesticide PQ^{2+} increased due to the presence of negatively charged sites in the semiconductor (Brigante & Avena, 2014). Figure 4.6 presents a schematic representation of paraquat adsorbed by SiO_2. According to Brigante et al., the increase in negative sites resulting from the loss of H^+ in the Si-OH groups can enhance the adsorption of PQ^{2+}.

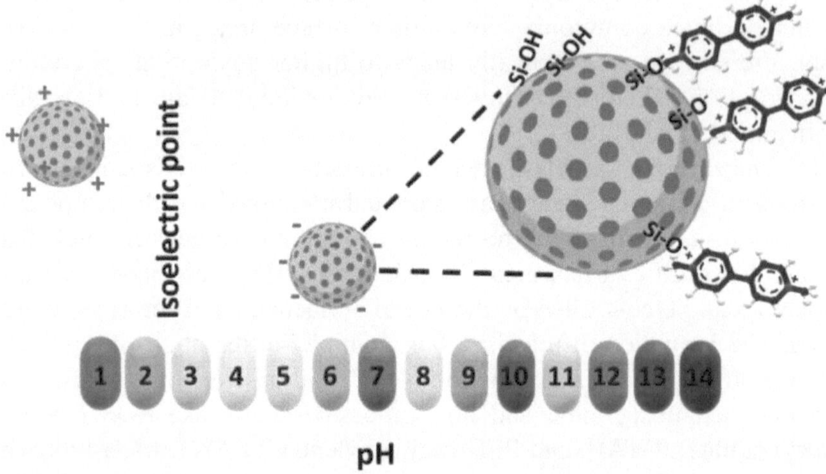

FIGURE 4.6 Adsorption of paraquat by SiO_2. Effect of the pH on the surface charge of the material.

A semiconductor can also be modified to change its charge surface. For example, Kaur et al. reported the functionalization of ZnO with CTAB (cetyltrimethylammonium bromide) and BMTF-IL (1-butyl-3-methyl-imidazolium tetrafluoroborate). The ZnO particles modified with CTAB presented a negatively charged surface at pH values higher than 7.64, while the ZnO modified with BMTF-IL presented a negatively charged surface at pH values higher than 5.67. The BMTF-IL functionalized ZnO showed a higher adsorption capacity in acid conditions compared to the CTAB functionalized ZnO (Kaur, Bhatia, Chaudhary, & Chaudhary, 2017).

4.2.2.3 CRYSTALLINE PROPERTIES

As mentioned above, the morphology of a material deeply affects the adsorption process. However, when using certain types of semiconductors like TiO_2, it is also important to pay attention to the crystal properties of the material. It has been proven that the adsorption of a compound on TiO_2 depends on the exposure of certain crystal facets, as different crystal facets can have different atomic surface structures (Liu, Yang, Pan, Yang, Lu, & Cheng, 2014).

Other properties related to the crystal structure, such as grain size or lattice strains, can also influence the adsorption process. A reduction in grain size and the production of structural defects like lattice strain create high energy sites that improve the adsorption capacity. These defects can be produced by the incorporation of dopants (Habibi, Rafiaei, Alhaji, & Zare, 2021).

Now that the main semiconductor properties that affect the adsorption process have been described, it is interesting to analyze some of the most common materials used in the adsorption of pesticides from water. Further section will provide an overview of these materials and their use for the removal of certain pesticides.

4.2.3 COMMON SEMICONDUCTOR ADSORBENTS FOR PESTICIDE IN WATER

It is well known that activated carbon is one of the most popular adsorbents for water pollutants (Amutova et al., 2023; Kumrić et al., 2019). However, there are some semiconductor materials that have proven to be good pesticide adsorbents. Titanium oxide (TiO_2) and silica (SiO_2) are two of the most common semiconductors used for pollutant removal from water. Other metal oxides and their composite materials have also been used for this purpose. Table 4.1 presents the adsorption of pesticides by some common semiconductors.

4.3 PHOTOCATALYTIC DEGRADATION OF PESTICIDES FOUND IN WATER

Another way of removing pesticides from water is through photocatalytic degradation. Photocatalysis is a process in which a semiconductor material is activated by light irradiation. If the energy of the light is higher than the bandgap of the semiconductor, the electrons in the valence band gain enough energy to move to the conduction band, creating a hole (+)

in the valence band. The resulting electron (–)/hole (+) pair reacts with the aqueous media in which the semiconductor is placed, forming free radicals that are responsible for degrading the pollutants (Coronado, 2013; Takahashi, Yoshikawa, & Sandhu, n.d.) (Figure 4.7).

TABLE 4.1 Common Semiconductor Adsorbents for Pesticide Removal from Water

Adsorbent	Pesticide	References
TiO_2	• Glyphosate • 4-Nitrophenol • AMPA • Methyl parathion	(Yang, Deng, & Zhang, 2019; Ilina et al., 2017; Allard et al., 2018)
SiO_2	• Chlorpyrifos • Carbendazim • Imidacloprid • Mesosulfuron-methyl • Paraquat	(Brigante & Avena, 2014; Wu et al., 2019; Ganiyu et al., 2014)
ZnO	• Naphthalene • Chlorpyrifos	(Kaur, Bhatia, Chaudhary, & Chaudhary, 2017; Gulati, Mandeep, Malik, & Kakkar, 2020)
Zeolites	• Glyphosate • Atrazine • Lindane • Diazinone	(Zavareh, Farrokhzad, & Darvishi, 2018; Lemić et al., 2006)

If the photocatalytic process is completed, the original organic compound can be fully degraded into small molecules such as benzene, CO_2, or H_2O, as seen in Figure 4.8. However, one of the main disadvantages of photocatalysis is the incomplete degradation of compounds, resulting in the generation of other toxic compounds. A solution to this problem is the combination of photocatalysis with an adsorption process.

4.3.1 PROPERTIES OF A SEMICONDUCTOR THAT AFFECTS THE PHOTOCATALYTIC PROCESS

As mentioned above, some authors report an adsorption process prior to the irradiation of the sample with light (Cardona, Węgrzyn,

Miskowiec, Ñorili, & Gil, 2023; Paul, Martens, & Frost, 2012). This is because in heterogeneous catalysis, reactions can occur on the surface of the material (Maron & Prutton, 1975). Therefore, morphology plays a crucial role in the photocatalytic process. Additionally, the morphology of the material can also affect the absorption of light. For instance, a highly porous morphology can promote light scattering and enhance the absorption of visible light (Estrada-Flores et al., 2020; Zhang et al., 2017).

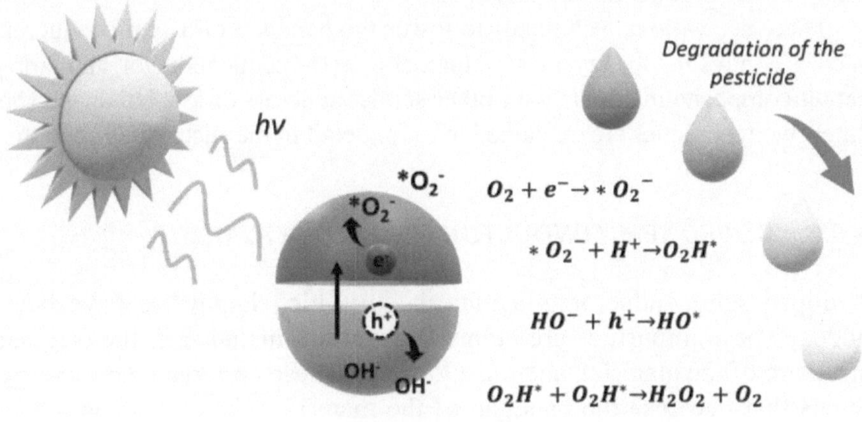

FIGURE 4.7 Activation of a semiconductor with light in a photocatalytic process.

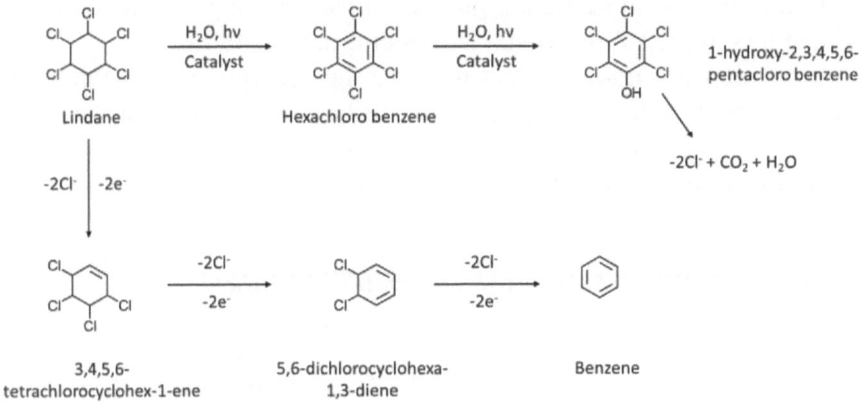

FIGURE 4.8 Degradation pathway of lindane.

Source: Proposed by Vaya & Surolia (2020); and Elliot, Lien, & Zhang (2009).

While morphology influences light absorption, the main property that affects the absorption of light by a semiconductor is the band-gap value. Semiconductors with a band-gap higher than 3.2 eV require high-energy light, typically in the UV range, to be activated. On the other hand, semiconductors with a band-gap lower than 3.2 eV can be activated with visible light. Common semiconductors like TiO_2 or ZnO only absorb light in the UV region of the electromagnetic spectrum due to their band-gap value. Therefore, significant efforts have been made to lower the bandgap of these materials, making them suitable for use under visible light (Lin & Yang, 2013; Macías-Sánchez et al., 2015).

There are various techniques to lower the bandgap of a semiconductor, such as controlling the crystal structure, doping the semiconductor, and fabricating composite materials with other semiconductors of low bandgap. The latter two techniques are explained in more detail in the succeeding sections.

4.3.2 DOPED SEMICONDUCTORS IN PHOTOCATALYSIS

A doped semiconductor is a material to which impurities have been added. These impurities are atoms that are not included in the original structure of the material, and the addition of these can create new energy levels that decrease the bandgap of the material. The atoms added can be metallic or non-metallic, and their incorporation affects not only the optical properties of the material but also its crystalline structure (Estrada-Flores et al., 2022). Figure 4.9 presents a doped semiconductor with atoms in the substitutional and interstitial positions.

Since doped semiconductors improve the absorption of visible light, the use of these materials in photocatalysis has become of great interest in the last couple of years. Table 4.2 shows the use of some doped semiconductor materials in the degradation of pesticides.

4.3.3 COMPOSITE MATERIALS IN PHOTOCATALYSIS

Another method to achieve the absorption of visible light by a semiconductor is the formation of composite materials. A composite material consists of two or more materials of different nature that are chemically united to create a material with improved properties. In photocatalysis, the formation of composite materials with semiconductors with different band-gap values is very useful because it can decrease the recombination rate of the electron (–)/hole (+) pair generated. The electrons in the conduction band of one of the semiconductors move to the conduction band of the other semiconductor instead of returning to the valence band (Yadav, Kumar, Kumari, Mittal, & Sharma, 2019; Soltani-Nezhad, Saljooqi, Mostafavi, & Shamspur, 2020). As a result, the process of photocatalysis continues for larger amounts of time. In Figure 4.10, it is possible to observe the mechanism of photocatalysis when a composite material is used.

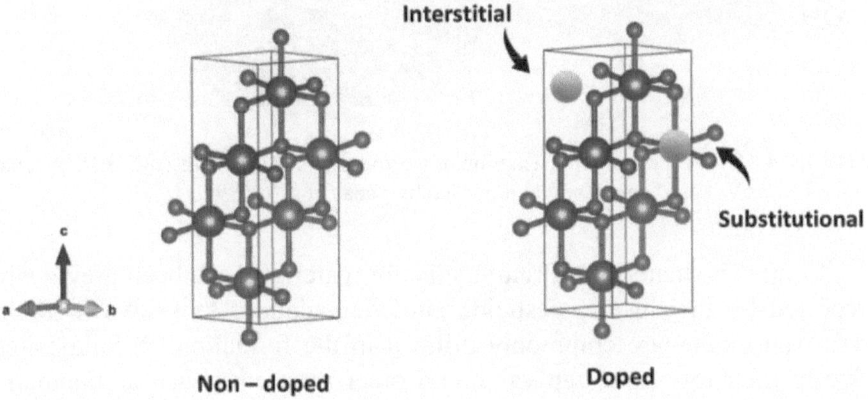

FIGURE 4.9 Crystal structure of a non-doped and a doped semiconductor made with the software VESTA (Momma & Izumi, 2011).

TABLE 4.2 Doped Semiconductor Materials Used for the Photodegradation of Pesticides

Semiconductor	Dopant	Photodegraded Pesticide	References
TiO_2	Ag	2-chlorophenol	Onkani et al. (2020)
	C, N	Diazinon	Boruah & Das (2020)
	Ni	Malathion	Surendra et al. (2020)
ZnO	Ag	2-chlorophenol	Onkani et al. (2020)
	N	Picloram	Macías-Sánchez et al. (2015)
	Cu	Monocrotophos	Hanh et al. (2019)
ZnS	Ag	Monocrotophos Chlorpyrifos	Ayodhya & Veerabhadram (2018)
Bi_2O_3	La	Chlorpyrifos	Khairnar et al. (2021)
Iron oxides	Bi	Metalaxyl	Gajović, Silva, Segundo, Šturm, Jančar, & Čeh (2011)

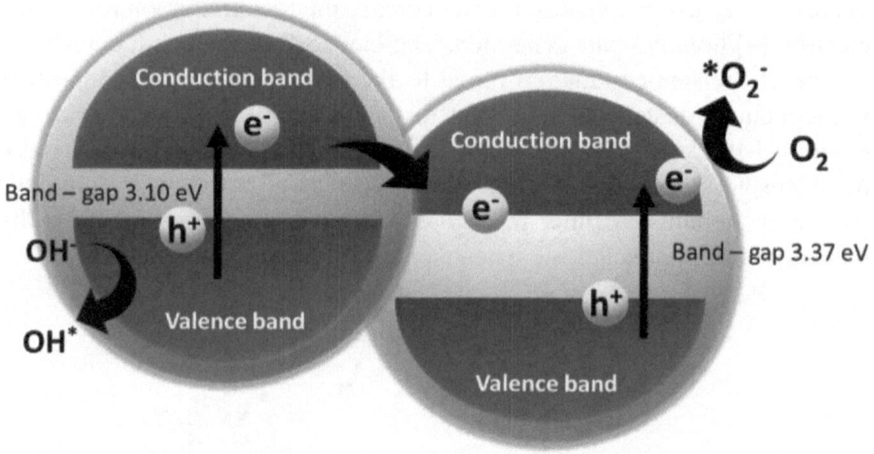

FIGURE 4.10 Schematic representation of a composite material with ZnO (3.10 eV) and SnO_2 (3.37 eV) based on the model proposed by Yadav et al. (2019).

Composite materials of semiconductor materials have been previously reported for their use in pesticide photodegradation. As shown in Table 4.2, iron oxides are commonly utilized in the formation of composites due to their low band-gap value and other properties such as magnetic behavior. The recombination of the electron/hole pair, which is commonly promoted by the low bandgap of the iron oxides, has been found to be inhibited with the formation of composites (Mirmasoomi, Mehdipour, & Galedari, 2017). These types of composite materials, along with other examples, are listed in Table 4.3.

4.4 CONCLUSIONS

With the increasing water contamination, the water treatment process represents an important area of opportunity. Although the removal of pesticides from water has been reported before, the common materials used for this purpose need to be improved. Techniques like adsorption or photocatalysis are being widely examined for the removal of pollutants from water. However, the design of the materials used in these applications needs to be carefully investigated, as it has been proven that the morphology, texture, optical, and structural properties of semiconductors influence the process. For that reason, research must be done in the field of new materials design for wastewater treatment and removal of toxic compounds from drinking water.

TABLE 4.3 Composite Semiconductor Materials Used for the Photodegradation of Different Pesticides

Composite Material	Type of Light	Photodegraded Pesticide	References
Fe_2O_3/TiO_2	UV	Triclosan Dichlorophenol Bisphenol	Cardona, Węgrzyn, Miskowiec, Ñorili, & Gil (2023)
ZnO/SnO_2	UV	Triclopyr	Yadav, Kumar, Kumari, Mittal, & Sharma (2019)
Fe_3O_4–TiO_2/graphene	Sunlight Visible UV	Atrazine	Boruah & Das (2020)
TiO_2/Fe_2O_3	Visible	Diazinon	Mirmasoomi, Mehdipour, & Galedari (2017)
Fe_3O_4/Cds-ZnS	Visible	Chlorpyrifos	Soltani-Nezhad, Saljooqi, Mostafavi, & Shamspur (2020)
TiO_2/SiO_2/carbon	UV	4-Nitrophenol	Wu et al. (2019)
GO-ZnO	UV	Quinalphos	Garg, Gupta, Singh, & Bansal (2022)
ZnO-Bi_2O_3	Visible	Cyhalothrin	Premalatha & Rose (2019)

KEYWORDS

- **adsorption**
- **composite materials**
- **crystalline properties**
- **isoelectric point**
- **pesticides**
- **photocatalytic process**
- **semiconductors**

REFERENCES

Allard, M. M., Merlos, S. N., Springer, B. N., Cooper, J., Zhang, G., Boskovic, D. S., Kwon, S. R., et al., (2018). Role of TiO_2 anatase surface morphology on organophosphorus interfacial chemistry. *J. Phys. Chem C., 122*, 29237–29248. doi: 10.1021/acs.jpcc.8b08641.

Amutova, F., Jurjanz, S., Akhmetsadykov, N., Kazankapova, M., Razafitianamaharavo, A., Renard, A., Nurseitova, M., et al., (2023). Adsorption of organochlorinated pesticides: Adsorption kinetic and adsorption isotherm study. *Results Eng., 17*. doi: 10.1016/j.rineng.2022.100823.

Ayodhya, D., & Veerabhadram, G., (2018). Ternary semiconductor ZnxAg1−xS nanocomposites for efficient photocatalytic degradation of organophosphorus pesticides. *Photochem. Photobiol. Sci., 17*, 1429–1442. doi: 10.1039/c8pp00220g. Cited: in: PMID: 30250952.

Bano, K., Kaushal, S., & Singh, P. P., (2021). A review on photocatalytic degradation of hazardous pesticides using heterojunctions. *Polyhedron [Internet], 209*, 115465. doi: 10.1016/j.poly.2021.115465.

Boruah, P. K., & Das, M. R., (2020). Dual responsive magnetic Fe_3O_4-TiO_2/graphene nanocomposite as an artificial nanozyme for the colorimetric detection and photodegradation of pesticide in an aqueous medium. *J. Hazard Mater. [Internet], 385*, 121516. doi: 10.1016/j.jhazmat.2019.121516. Cited: in: PMID: 31708291.

Brigante, M., & Avena, M., (2014). Synthesis, characterization and application of a hexagonal mesoporous silica for pesticide removal from aqueous solution. *Microporous Mesoporous Mater. [Internet], 191*, 1–9. doi: 10.1016/j.micromeso.2014.02.035.

Cardona, Y., Węgrzyn, A., Miskowiec, P., Ñorili, S. A., & Gil, A., (2023). Heterogeneous Fenton- and photo-Fenton-like catalytic degradation of emerging pollutants using Fe_2O_3/TiO_2/pillared clays synthesized from aluminum industrial wastes. *J. Water Process Eng., 52*, 103494. doi: https://doi.org/10.1016/j.jwpe.2023.103494.

Chen, K., Li, J., Wang, W., Zhang, Y., Wang, X., & Su, H., (2012). Effects of surfactants on microstructure and photocatalytic activity of TiO_2 nanoparticles prepared by the hydrothermal method. *Mater. Sci. Semicond. Process [Internet], 15*, 20–26. doi: 10.1016/j.mssp.2011.05.007.

Coronado, J. M., (2013). Photons, electrons and holes: Fundamentals of photocatalysis with semiconductors. *Des. Adv. Photocatalytic Mater. Energy Environ. Appl*. Springer.

Data of the Food and Agriculture Organization of the United Nations [Internet], (2020). Pestic. Use. Available from: https://www.fao.org/faostat/en/#data/RP (accessed on 24 February 2024).

Elliott, D. W., Lien, H. L., & Zhang, W. X., (2009). Degradation of lindane by zero-valent iron nanoparticles. *J. Environ. Eng., 135*, 317–324. doi: 10.1061/(asce)0733-9372(2009)135: 5(317).

El-Nahhal, I., & El-Nahhal, Y., (2021). Pesticide residues in drinking water, their potential risk to human health and removal options. *J. Environ. Manage [Internet], 299*, 113611. doi: 10.1016/j.jenvman.2021.113611. Cited: in: PMID: 34526283.

Estrada-Flores, S., Martínez-Luévanos, A., Perez-Berumen, C. M., García-Cerda, L. A., & Flores-Guia, T. E., (2020). Relationship between morphology, porosity, and the photocatalytic activity of TiO_2 obtained by sol–gel method assisted with ionic and nonionic surfactants. *Bol. la Soc. Esp Ceram y Vidr., 59*, 209–218. doi: 10.1016/j.bsecv.2019.10.003.

Estrada-Flores, S., Pérez-Berumen, C. M., Flores-Guia, T. E., García-Cerda, L. A., Rodríguez-Hernández, J., Esquivel-Castro, T. A., & Martínez-Luévanos, A., (2022). Mechanosynthesis of mesoporous Bi-doped TiO_2: The effect of bismuth doping and ball milling on the crystal structure, optical properties, and photocatalytic activity. *Crystals., 12*.

Gajović, A., Silva, A. M. T., Segundo, R. A., Šturm, S., Jančar, B., & Čeh, M., (2011). Tailoring the phase composition and morphology of Bi-doped goethite-hematite nanostructures and their catalytic activity in the degradation of an actual pesticide using a photo-Fenton-like process. *Appl. Catal. B Environ., 103*, 351–361. doi: 10.1016/j.apcatb.2011.01.042.

Ganiyu, S. O., Bispo, C., Bion, N., Ferreira, P., & Batonneau-Gener, I., (2014). Periodic mesoporous organosilicas as adsorbents for the organic pollutants removal in aqueous phase. Microporous *Mesoporous Mater. [Internet], 200*, 117–123. doi: 10.1016/j.micromeso.2014.07.047.

Garg, R., Gupta, R., Singh, N., & Bansal, A., (2022). Eliminating pesticide quinalphos from surface waters using synthesized GO-ZnO nanoflowers: Characterization, degradation pathways and kinetic study. *Chemosphere [Internet], 286*, 131837. doi: 10.1016/j.chemosphere.2021.131837. Cited: in: PMID: 34399266.

Gerasimova, T. V., Evdokimova (Galkina), O. L., Kraev, A. S., Ivanov, V. K., & Agafonov, A. V., (2016). Micro-mesoporous anatase TiO_2 nanorods with high specific surface area possessing enhanced adsorption ability and photocatalytic activity [Internet]. *Microporous Mesoporous Mater*. Elsevier Ltd. Available from: http://dx.doi.org/10.1016/j.micromeso.2016.08.015.

Gulati, A., Mandeep, Malik, J., & Kakkar, R., (2020). Mesoporous rGO@ZnO composite: Facile synthesis and excellent water treatment performance by pesticide adsorption and catalytic oxidative dye degradation. *Chem. Eng. Res. Des. [Internet], 160*, 254–263. doi: 10.1016/j.cherd.2020.04.040.

Habibi, M. K., Rafiaei, S. M., Alhaji, A., & Zare, M., (2021). $ZnAl_2O_4$: Ce^{3+} phosphors: Study of crystal structure, microstructure, photoluminescence properties and efficient adsorption of Congo red dye. *J. Mol Struct., 1228*. doi: 10.1016/j.molstruc.2020.129769.

Hanh, N. T., Le Minh, T. N., Van, T. D., Thanh, T. M. H., Pham, T. D., Minh, T. D., Trang, H. T., Binh, M. T., & Nguyen, M. V., (2019). Monocrotophos pesticide effectively removed by novel visible light driven Cu doped ZnO photocatalyst. *J. Photochem. Photobiol. A Chem. [Internet], 382*, 111923. doi: 10.1016/j.jphotochem.2019.111923.

Hitam, C. N. C., & Jalil, A. A., (2020). A review on exploration of Fe_2O_3 photocatalyst towards degradation of dyes and organic contaminants. *J. Environ. Manage [Internet], 258*, 110050. doi: 10.1016/j.jenvman.2019.110050. Cited: in: PMID: 31929077.

Ilina, S. M., Ollivier, P., Slomberg, D., Baran, N., Pariat, A., Devau, N., Sani-Kast, N., et al., (2017). Investigations into titanium dioxide nanoparticle and pesticide interactions in aqueous environments. *Environ. Sci. Nano, 4*, 2055–2065. doi: 10.1039/c7en00445a.

Kaur, Y., Bhatia, Y., Chaudhary, S., & Chaudhary, G. R., (2017). Comparative performance of bare and functionalize ZnO nanoadsorbents for pesticide removal from aqueous solution. J. *Mol. Liq. [Internet], 234*, 94–103. doi: 10.1016/j.molliq.2017.03.069.

Khairnar, S. D., Kulkarni, A. N., Shinde, S. G., Marathe, S. D., Marathe, Y. V., Dhole, S. D., & Shrivastava, V. S., (2021). Synthesis and characterization of 2-D La-doped Bi_2O_3 for photocatalytic degradation of organic dye and pesticide. *J. Photochem. Photobiol., 6*, 100030. doi: 10.1016/j.jpap.2021.100030.

Kumrić, K., Vujasin, R., Egerić, M., Petrović, Đ., Devečerski, A., & Matović, L., (2019). Coconut shell activated carbon as solid-phase extraction adsorbent for preconcentration of selected pesticides from water samples. *Water Air Soil Pollut., 230*. doi: 10.1007/s11270-019-4359-7.

Lemić, J., Kovačević, D., Tomašević-Čanović, M., Kovačević, D., Stanić, T., & Pfend, R., (2006). Removal of atrazine, lindane and diazinone from water by organo-zeolites. *Water Res., 40,* 1079–1085. doi: 10.1016/j.watres.2006.01.001. Cited: in: PMID: 16460782.

Li, X., He, G., Xiao, G., Liu, H., & Wang, M., (2009). Synthesis and morphology control of ZnO nanostructures in microemulsions. *J. Colloid Interface Sci. [Internet], 333*, 465–473. doi: 10.1016/j.jcis.2009.02.029. Cited: in: PMID: 19286190.

Lin, C. J., & Yang, W. T., (2013). Ordered mesostructured Cu-doped TiO_2 spheres as active visible-light-driven photocatalysts for degradation of paracetamol. *Chem. Eng. J. [Internet], 237*, 131–137. doi: 10.1016/j.cej.2013.10.027.

Liu, G., Yang, H. G., Pan, J., Yang, Y. Q., Lu, G. Q. M., & Cheng, H. M., (2014). Titanium dioxide crystals with tailored facets. *Chem. Rev., 114*, 9559–9612. doi: 10.1021/cr400621z. Cited: in: PMID: 24851995.

Macías-Sánchez, J. J., Hinojosa-Reyes, L., Caballero-Quintero, A., De La Cruz, W., Ruiz-Ruiz, E., Hernández-Ramírez, A., & Guzmán-Mar, J. L., (2015). Synthesis of nitrogen-doped ZnO by sol-gel method: Characterization and its application on visible photocatalytic degradation of 2,4-D and picloram herbicides. *Photochem. Photobiol. Sci., 14*, 536–542. doi: 10.1039/c4pp00273c. Cited: in: PMID: 25337907.

Maron, S. H., & Prutton, C. F., (1975). Surface phenomenon and catalysis. *Princ. Phys. Chem.* (pp. 813–847). The McMillan Publishing Co.

Mirmasoomi, S. R., Mehdipour, G. M., & Galedari, M., (2017). Photocatalytic degradation of diazinon under visible light using TiO_2/Fe_2O_3 nanocomposite synthesized by ultrasonic-assisted impregnation method. *Sep. Purif. Technol. [Internet], 175*, 418–427. doi: 10.1016/j.seppur.2016.11.021.

Mohammed, A. A., & Ali, D. K., (2023). Bentonite-layered double hydroxide composite as potential adsorbent for removal of abamectin pesticide from wastewater. *Results in Surfaces and Interfaces [Internet], 10,* 100099. doi: 10.1016/j.rsurfi.2023.100099.

Momma, K., & Izumi, F., (2011). VESTA 3 for three-dimensional visualization of crystal, volumetric and morphology data. *J. Appl. Crystallogr., 44*, 1272–1276.

Nagamine, S., & Sasaoka, E., (2002). Synthesis of nanostructured titania templated by anionic surfactant in acidic conditions. *J. Porous Mater., 9*, 167–173. doi: 10.1023/A:1020966016542.

Onkani, S. P., Diagboya, P. N., Mtunzi, F. M., Klink, M. J., Olu-Owolabi, B. I., & Pakade, V., (2020). Comparative study of the photocatalytic degradation of 2–chlorophenol under UV irradiation using pristine and Ag-doped species of TiO_2, ZnO and ZnS photocatalysts. *J. Environ. Manage [Internet], 260*, 110145. doi: 10.1016/j.jenvman.2020.110145. Cited: in: PMID: 32090837.

Ore, O. T., Adeola, A. O., Bayode, A. A., Adedipe, D. T., & Nomngongo, P. N., (2023). Organophosphate pesticide residues in environmental and biological matrices: Occurrence, distribution, and potential remedial approaches. *Environ. Chem. Ecotoxicol. [Internet], 5*, 9–23. doi: 10.1016/j.enceco.2022.10.004.

Pan, G., Wang, D., & Liu, Y. Y., (2012). Photocatalytic degradation pathways and adsorption modes of H-acid in TiO_2 suspensions. *Chinese Sci Bull., 57*, 1102–1108. doi: 10.1007/s11434-011-4894-0.

Paul, B., Martens, W. N., & Frost, R. L., (2012). Immobilized anatase on clay mineral particles as a photocatalyst for herbicides degradation. *Appl. Clay Sci. [Internet], 57*, 49–54. doi: 10.1016/j.clay.2011.12.009.

Premalatha, N., & Rose, M. L., (2019). Surfactant modified ZnO–Bi_2O_3 nanocomposite for degradation of lambda–cyhalothrin pesticide in visible light: A study of reaction kinetics and intermediates. *J. Environ. Manage. [Internet], 246*, 259–266. doi: 10.1016/j.jenvman.2019.05.155. Cited: in: PMID: 31181474.

Sharma, M., Hazra, S., & Basu, S., (2017). Kinetic and isotherm studies on adsorption of toxic pollutants using porous ZnO@SiO_2 monolith. *J. Colloid Interface Sci. [Internet], 504*, 669–679. doi: 10.1016/j.jcis.2017.06.020. Cited: in: PMID: 28622560.

Soltani-Nezhad, F., Saljooqi, A., Mostafavi, A., & Shamspur, T., (2020). Synthesis of Fe_3O_4/CdS–ZnS nanostructure and its application for photocatalytic degradation of chlorpyrifos pesticide and brilliant green dye from aqueous solutions. *Ecotoxicol. Environ. Saf. [Internet], 189,* 109886. doi: 10.1016/j.ecoenv.2019.109886. Cited: in: PMID: 31759746.

Surendra, B., Raju, B. M., Noel, K., Onesimus, S., Choudhary, G. L., Paul, P. F., & Vangalapati, M., (2020). Characterization of Ni doped TiO_2 nanoparticles and its application for the degradation of malathion. *Mater Today Proc. [Internet], 26*, 1091–1095. doi: 10.1016/j.matpr.2020.02.216.

Sustainable Development Goals of the United Nations Development Program [Internet], (2015). Available from: https://www.undp.org/sustainable-development-goals?utm_source=EN&utm_medium=GSR&utm_content=US_UNDP_PaidSearch_Brand_English&utm_campaign=CENTRAL&c_src=CENTRAL&c_src2=GSR&gclid=Cj0KCQiA2-2eBhClARIsAGLQ2RnDvr4BNuOyzHeOABCreoJzJKn-gw9pxT115mJEKJwqeuktqCb9DN (accessed on 24 February 2024).

Takahashi, K., Yoshikawa, A., & Sandhu, A. (2014). *Wide Bandgap Semiconductors*. Springer.

Tho, P. T., Van, H. T., Nguyen, L. H., Hoang, T. K., Ha Tran, T. N., Nguyen, T. T., Hanh, N. T. B., Nguyen, V. Q., Le Sy, H., Thai, V. N., et al., (2021). Enhanced simultaneous adsorption of As(iii), Cd(ii), Pb(ii) and Cr(vi) ions from aqueous solution using cassava root husk-derived biochar loaded with ZnO nanoparticles. *RSC Adv., 11*, 18881–18897. doi: 10.1039/d1ra01599k.

Vaya, D., & Surolia, P. K., (2020). Semiconductor based photocatalytic degradation of pesticides: An overview. *Environ. Technol. Innov. [Internet], 20*, 101128. doi: 10.1016/j.eti.2020.101128.

Wei, Y., Xu, R. X., Gao, C., Liu, J. H., & Huang, X. J., (2012). Polishing-activated nano α-Al_2O_3: Adsorption and electrochemical behavior toward organophosphate pesticides. *Electrochem. Commun. [Internet], 18*, 78–80. doi: 10.1016/j.elecom.2012.02.007.

Wu, X. Q., Shao, Z. D., Liu, Q., Xie, Z., Zhao, F., & Zheng, Y. M., (2019). Flexible and porous TiO_2/SiO_2/carbon composite electrospun nanofiber mat with enhanced interfacial charge separation for photocatalytic degradation of organic pollutants in water. *J. Colloid Interface Sci. [Internet], 553*, 156–166. doi: 10.1016/j.jcis.2019.06.019. Cited: in: PMID: 31202052.

Yadav, S., Kumar, N., Kumari, V., Mittal, A., & Sharma, S., (2019). Photocatalytic degradation of triclopyr, a persistent pesticide by ZnO/SnO_2 nano-composities. *Mater. Today Proc. [Internet], 19,* 642–645. doi: 10.1016/j.matpr.2019.07.746.

Yang, Y., Deng, Q., & Zhang, Y., (2019). Comparative study of low-index {1 0 1}-TiO_2, {0 0 1}-TiO_2, {1 0 0}-TiO_2 and high-index {2 0 1}-TiO_2 on glyphosate adsorption and photo-degradation. *Chem Eng J., 360,* 1247–1254. doi: 10.1016/j.cej.2018.10.219.

Yeganeh, M., Charkhloo, E., Sobhi, H. R., Esrafili, A., & Gholami, M., (2022). Photocatalytic processes associated with degradation of pesticides in aqueous solutions: Systematic review and meta-analysis. *Chem. Eng. J. [Internet], 428,* 130081. doi: 10.1016/j.cej.2021.130081.

Zavareh, S., Farrokhzad, Z., & Darvishi, F., (2018). Modification of zeolite 4A for use as an adsorbent for glyphosate and as an antibacterial agent for water. *Ecotoxicol. Environ. Saf. [Internet], 155,* 1–8. doi: 10.1016/j.ecoenv.2018.02.043. Cited: in: PMID: 29486406.

Zhang, G., Zhang, S., Wang, L., Liu, R., Zeng, Y., Xia, X., Liu, Y., & Luo, S., (2017). Facile synthesis of bird's nest-like TiO_2 microstructure with exposed (001) facets for photocatalytic degradation of methylene blue. *Appl. Surf. Sci. [Internet], 391,* 228–235. doi: 10.1016/j.apsusc.2016.04.095.

Zhou, Q., & Fang, Z., (2015). Graphene-modified TiO_2 nanotube arrays as an adsorbent in micro-solid phase extraction for determination of carbamate pesticides in water samples. *Anal. Chim. Acta [Internet], 869,* 43–49. doi: 10.1016/j.aca.2015.02.019. Cited: In: PMID: 25818138.

CHAPTER 5

Microencapsulation of Bioactive Compounds as a Sustainable Option for Treatment of Inflammatory Conditions

J. D. GARCÍA-ORTÍZ,[1] ADRIANA CAROLINA FLORES-GALLEGOS,[1] JUAN ALBERTO ASCACIO-VALDÉS,[1] C. N. AGUILAR-GONZÁLEZ,[1] S. D. FLORES-NERY,[1] S. C. ESPARZA-GONZÁLES,[2] and RAÚL RODRÍGUEZ-HERRERA[1]

[1]*School of Chemistry, Universidad Autónoma de Coahuila, Blvd. Venustiano Carranza e Ing. José Cárdenas V. s/n. Col., República Oriente, Saltillo, Coahuila, México*

[2]*School of Dentistry-Saltillo, Universidad Autónoma de Coahuila, Avenida Dra, Cuquita Cepeda de Dávila s/n. Col., Adolfo López Mateos, Saltillo, Coahuila, México*

ABSTRACT

Bioactive compounds have a wide variety of activities or effects that are of interest to human health. Unfortunately, a large amount of the foods that are consumed, despite containing a high content of bioactive compounds, lose their functions after being subjected to digestive processes or different environmental conditions. This limits or eliminates their potential. Microencapsulated products have a great advantage in this regard, as they possess a large number of different characteristics, with stability in various environments being a standout feature. The following document presents a review of the different types of processes to encapsulate bioactive

Sustainable Environment and Health: Practical Strategies. Leonardo Sepúlveda Torre, Ann Rose Abraham, Neha Kanwar Rawat, & A. K. Haghi (Eds.)

compounds, along with necessary considerations for their formulation, advantages of their bioavailability, and applications as an alternative to treat problems related to inflammation.

5.1 INTRODUCTION

Worldwide, the main cause of human death is due to chronic-degenerative diseases (WHO, 2021), which is of great concern and generates interest in the search for treatments and methods to prevent these diseases. Diet and lifestyle are involved in these types of diseases since they are directly related to oxidative stress. Some changes in diets can attenuate or reduce symptoms or functional deterioration caused by these diseases. Different studies in nutrigenomics have documented a complex relationship between diseases, diet, and genetics (Tumani et al., 2020; Rodríguez-Graña et al., 2015; Yang et al., 2021).

Different diets with a high content of bioactive compounds have been tried to reduce symptom progression of various diseases such as muscular dystrophy, arthritis, pulmonary emphysema, amyloidosis, collagenosis, Friedreich's ataxia, lipofuscinosis, peptic ulcer, inflammatory bowel diseases, Crohn's disease, and even different cancers (Rodríguez-Graña et al., 2015; Yang et al., 2021). This is because diet has recently become an alternative for both treatment and prevention due to the multiple effects that this type of diet can generate in the body (Chhikara et al., 2018).

Different bioactive compounds, even when present in low concentrations, have been reported to have significant therapeutic activity for certain pathologies (Chhikara et al., 2018; Konstantinidi & Koutelidakis, 2019; Zhang et al., 2020).

The use of bioactive compounds from natural origins, such as polyphenols, to treat pathologies may offer advantages since they usually have low toxicity or minor side effects compared to conventional drugs (Lee et al., 2017). Additionally, some polyphenols, such as anthocyanins and phenolic acids, have been found to reduce inflammatory processes, attenuating one of the main symptoms and complications during inflammatory diseases (Tumani et al., 2020; Rodríguez Graña et al., 2015; Yang et al., 2021).

However, one of the most important disadvantages of this type of compound is their low stability, which makes them susceptible to physical-chemical changes, causing them to lose their bioactive properties and reducing their effects on human health, among other factors. Some of the factors that can affect polyphenol stability include light, temperature, and conditions after polyphenol consumption, especially those that may occur in the digestive tract, where polyphenols are partially absorbed, and their beneficial effects are diminished.

Microencapsulation is one strategy to increase polyphenol stability, which involves depositing these components in a matrix similar to protective walls, providing greater stability and increasing their lifespan under stress conditions such as variations in pH, temperature, light, oxidative stress, and others. Furthermore, microencapsulation can increase gastrointestinal retention time and enhance bio-adhesive properties in the intestinal epithelium (Lee et al., 2017; Tarone et al., 2020).

5.2 BIOACTIVE COMPOUNDS

Chronic degenerative diseases are primarily influenced by people's diet and lifestyle. Certain food ingredients and additives are associated with oxidative stress, which can lead to clinical complications such as inflammation or cancer. While specific treatments for conditions like inflammatory bowel disease or Crohn's disease are not yet available, studies have shown that inflammation can be reduced through diets rich in bioactive compounds with phenolic properties, such as anthocyanins and phenolic acids (Tumani et al., 2020; Rodríguez Graña et al., 2015; Yang et al., 2021).

Bioactive compounds have the potential to be used as treatments or adjuvant treatments for various pathologies due to their hypoglycemic, anti-inflammatory, anti-anemic, antibacterial, antioxidant, antiallergic, hepatoprotective, hypolipidemic, and antipyretic activities (Chhikara et al., 2018). Additionally, these compounds can impart different physical-chemical characteristics and sensory properties to food matrices, enhancing their appearance and resulting in greater acceptance and demand by society. Moreover, the consumption of high-bioactive compounds food may have positive effects on health (Abdel-Aal et al., 2006; Bobek et al., 2001; El-Baz et al., 2020; Sharma et al., 2022; von Lintig et al., 2020).

Currently, numerous bioactive compounds have been identified, including polypeptides, prebiotics, and phenolic compounds such as carotenoids, flavonoids, and sterols (Martin et al., 2014; Sepúlveda-Vázquez et al., 2018).

5.3 MICROENCAPSULATION

5.3.1 MICROENCAPSULATION OF BIOACTIVE COMPOUNDS

Due to the low stability of bioactive compounds, their beneficial activity is limited after consumption. Therefore, the search for strategies to enhance their stability against factors such as light, temperature, oxidation, pH, etc., is a current topic of interest. One widely used method to maintain the bioactive properties and stability of different compounds is microencapsulation. This technique involves creating functional barriers at microscopic levels, forming spheres or capsules (Figure 5.1) of various sizes, ranging from one micron to hundreds, depending on the specific technique and method employed. Microencapsulation serves the purpose of protecting the encapsulated material and preserving its biological, functional, and physicochemical properties for a certain period of time. It allows for either total or partial release, achieving a gradual and prolonged release, which is known as diffusion or controlled release (Kaur et al., 2021; Sri et al., 2013; Suganya & Anuradha, 2017; Tarone et al., 2020).

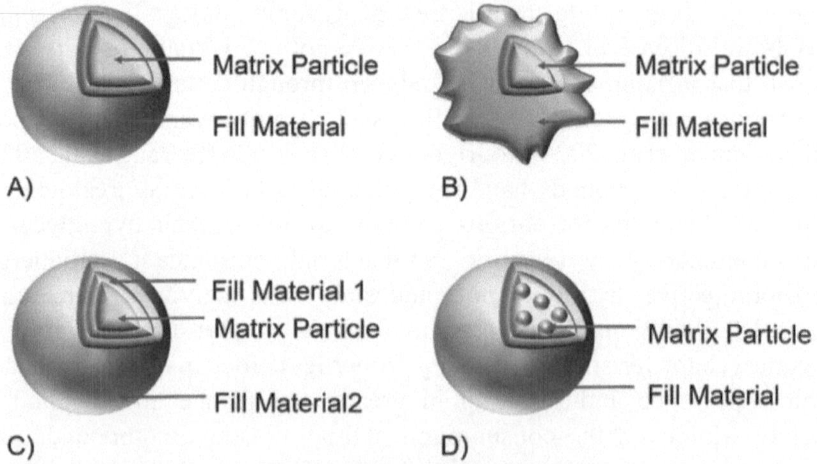

FIGURE 5.1 Types of microencapsulation – (A) Regular encapsulation with a single encapsulation material and a single encapsulated matrix; (B) irregular encapsulation with a single encapsulation material and a single encapsulated matrix; (C) regular encapsulation with multiple encapsulation materials and a single encapsulated matrix. And single encapsulated matrix – (D) regular encapsulation with a single encapsulating material and multiple encapsulated matrices.

Source: Prepared by the author.

In this process, there are different factors such as the concentration of the surrounding matrix, its solubility, and the elimination of the solvent, which are of utmost importance. The purpose of microencapsulation is to produce materials wrapped in others, providing them with greater stability. For microencapsulation, different techniques and materials can be used, resulting in different morphologies and release mechanisms depending on the objectives, benefits, and fields of application (Sri et al., 2013).

Recently, microencapsulation has been recognized as one of the most viable methods to provide greater stability to bioactive compounds (Lee et al., 2017). This method confers better characteristics to the encapsulated compounds, improving their conservation, stability, and resistance to different environmental factors that would otherwise cause them to lose their activity (Oancea et al., 2018). Microencapsulation enhances the chemical stability of various bioactive compounds, which can have beneficial effects on health by prolonging storage and antioxidant activity (Chutia & Mahanta, 2021). Furthermore, the incorporation of different chemical compounds with biological activities can increase stability and retention in these matrices (Chuyen et al., 2019; Ozkan et al., 2019), further improving characteristics such as resistance to degradation and factors like heat, radiation, or gastrointestinal conditions (Yan et al., 2021), which play a crucial role in the degradation of these compounds.

5.3.2 TECHNIQUES

Table 5.1 shows different processes for obtaining microencapsulation as well as its fundaments.

Each of the methods has different advantages and disadvantages. The selection of method, as well as materials to be encapsulated, may depend on equipment accessibility, type of material to be encapsulated, cover materials, and desired characteristics in microencapsulation (Choudhury et al., 2021; Niño-Vásquez et al., 2022).

5.3.3 COATING MATERIALS

The materials used for microencapsulation vary according to the required traits. In the core, which is the material intended to be encapsulated, different characteristics and properties may be present, such as stabilizers, diluents, excipients, and release rate retarders or accelerators. There are two main types of materials: hydrophilic and hydrophobic. The selection of these will depend on the intended function or the effect that is intended to be studied.

TABLE 5.1 Main Microencapsulation Techniques

Technique	Fundament	References
Emulsification	Production of small diameter wet microcapsules formed by the interaction of oil or some organic material with water.	Richards & Malik (2021); Zhang et al. (2020)
Spray dry	It consists of forming an encapsulated powder form from a liquid or emulsion which is heated by an air current between 100°C and 300°C.	Mohammed et al. (2020); Trang Nguyen et al. (2021)
Spray cooling	It consists of forming an encapsulated solid from a liquid or emulsion which is cooled using a chamber of cold air or liquid nitrogen.	Choudhury et al. (2021); Favaro-Trindade et al. (2021)
Electrospray	Solid particles are produced from an air stream where the encapsulation is coated with a molten polymer.	Choudhury et al. (2021)
Lyophilization	Encapsulation occurs with rapid freezing and removal of ice by pressure change.	Otálora et al. (2022); Rodrigues et al. (2021)
Coacervation	It is produced by phase separation of colloidal systems.	Castro-Rosas et al. (2017)
Polymerization	It is based on the polymerization of a monomer forming a shell around a component.	Revuelta et al. (2021)
Extrusion	A film is formed by interaction of a polysaccharide with an iron retaining the nucleus.	Ribeiro & Veloso (2021)
Supercritical fluids	Impregnation is carried out at temperatures close to room temperature using supercritical carbon dioxide.	Gañan et al. (2020)
Co-crystallization	It is produced through crystallization of carbohydrates that trap the materials to be encapsulated.	Ribeiro & Veloso (2021)

Materials used to form the coating must have the ability to form a cohesive film with the encapsulated material. They must show chemical compatibility and not react with any other material used for encapsulation or the core, as this could limit bioactivity or coating characteristics. In addition, it is sought that coating materials provide properties such as resistance, flexibility, impermeability, optical properties, and stability.

The material used to form the wall or microcapsule should promote stabilization of the core material. It should be inert towards active ingredients and control release under specific conditions. Furthermore, it should be film-forming, flexible, tasteless, without high viscosity, inexpensive, stable, non-hygroscopic, and soluble in an aqueous or solvent medium. In addition, the melting and coating must be flexible, brittle, hard, and thin (Maroulas et al., 2009; Paulo & Santos, 2017; Suganya & Anuradha, 2017). During microcapsule formulation, at least two components are required: the compound that will be encapsulated and the encapsulating shell. The compounds used for microencapsulation should be safe, stable, biocompatible, and have a predictable degradation process (Paulo & Santos, 2017). Coating composition can vary according to the biological materials used, such as proteins, wax, carbohydrates, fats, polymers, or copolymers, which may have natural or synthetic origins (Nahum & Domb, 2021; Nesterenko et al., 2013).

The use of materials of biological origin has shown advantages such as greater biocompatibility, biodegradability, good amphiphilic and functional properties (water solubility, emulsifying and foaming capacity), making them an attractive option for the pharmaceutical, cosmetic, and food industries (Nesterenko et al., 2013). Additionally, natural polymers offer benefits such as biocompatibility, biodegradability, and good amphiphilic and functional properties (water solubility, emulsifying and foaming capacity). The use of plant proteins as wall-forming materials in microencapsulation reflects the current "green" trend in these industries. In food applications, plant proteins are known to be less allergenic compared to animal proteins (Bowen et al., 2019; Nesterenko et al., 2013). The most widely used wall materials include proteins, carbohydrates, gums, lipids, various emulsifiers, and fibers, either alone or in combination with other compounds (Table 5.2).

Among the most commonly used components for the successful production of lipid-based emulsion or gel microencapsulates are whey protein, starch, gelatin, casein, and zein (Wang et al., 2020).

Microencapsulates are a viable option as an aid in various treatments, especially for anti-inflammatories that are required in locations beyond the digestive process. These microencapsulates can be formed from two or more materials, combining their properties. The materials used should be low-cost and the techniques employed for their production should be environmentally friendly and sustainable (Gonzalez-Reza et al., 2018; Gutierrez-Alvarado et al., 2022).

TABLE 5.2 Different Encapsulating Materials, Properties, and Techniques (Choudhury et al., 2021; Kashappa & Hyun, 2005)

Fill Material	Source	Properties	Techniques Used
Gum	Arabic gum, sodium alginate.	Forms soft elastic gels, low tensile strength, hydrocolloid.	Extrusion, phase separation Spray drying, coacervation, emulsification.
Carbohydrates	Starch, dextran, sucrose.	Hydrocolloid, comparatively higher tensile strength than rubbers.	Spray drying, fluidized bed coating, extrusion, lyophilization.
Protein	Gelatin, albumin	Emulsification, gelation, foam formation and water retention capacity.	Spray drying, extrusion, coacervation, emulsification by lyophilization.
Lipids	Beeswax, stearic acid, phospholipids.	Plasticizing properties, good barrier to gases and water vapors.	Fluidized bed coating, cooling/spray cooling, extrusion.
Cellulose and its derivatives	Vegetable cells	Hydrophilic, good film-forming ability and surface activity.	Spray drying, fluidized bed coating, extrusion, emulsification/precipitation coacervation.
Chitosan	Crustacean shells	Good barrier to gases and water vapors.	Spray drying, coacervation, emulsification.

5.4 BIOAVAILABILITY OF ENCAPSULATED COMPOUNDS

Normally, bioactive compounds have multiple beneficial effects that are directly influenced by their bioavailability, which refers to the speed and extent to which an active substance is absorbed and reaches its site of absorption or action (Teodora et al., 2019). Anthocyanins serve as a clear example, as they possess anti-inflammatory and anti-cancer properties. However, their bioavailability is often limited (Sharif et al., 2020), influenced by various factors such as temperature and pH changes, as well as enzymatic processes in the digestive and assimilation system. These factors can alter the bioavailability and distribution of anthocyanins (Grgić et al., 2020; Martin et al., 2014; Tena et al., 2020; Teodora et al., 2019).

Different steps during the digestion and assimilation processes of bioactive compounds in the gastrointestinal tract are influenced by various factors that play an important role in the bioavailability of the compounds and their activities (Teodora et al., 2019). These processes create conditions that affect the structure of the compounds, starting with mechanical crushing and enzymatic processes in the mouth, where the particle size is reduced. Then, the release of food components begins, and changes are made through interaction with saliva. Acid digestion takes place in the stomach and intestinal juices, where different enzymatic activities are initiated to break down compounds such as carbohydrates. As the digestive process continues, the size of the consumed food is further reduced, releasing more of its components and promoting their interaction with the environment. While some compounds are freely absorbed in the stomach, a significant number of compounds are not absorbed due to the environmental conditions of the stomach, where processes such as hydrolysis occur at a very low pH (2–4) (Grgić et al., 2020). Subsequently, the undigested compounds undergo intestinal conditions where the pH changes back to a neutral level. It has been documented that depending on the type of bioactive molecule, they can be retained in this section and released in the small intestine, while molecules that are still intact, as well as residues of other molecules, pass into the large intestine section (Guo et al., 2022).

Some microencapsulates have been shown to have good release under both acidic and neutral conditions, suggesting that they can be subjected to both stomach and intestinal digestion conditions (Araba et al., 2019; Stefanescu et al., 2022). When gastric and intestinal conditions were simulated, multiple coatings showed good potential for nutrient encapsulation, such as whey protein isolates, because of their resistance to gastric and intestinal conditions (Dehkordi et al., 2020). Additionally, some microencapsulates can be promising systems for the active principle, as they can exhibit high sensitivity to pH, allowing for the creation of targeted treatments (Braim et al., 2019; de Oliveira Pedro et al., 2018).

The aforementioned pH changes allow for the secretion and activation of different pancreatic and bile enzymes, which contribute to the absorption of bioactive components. Some *in vitro* models have shown that microencapsulation can support enzymatic processes, such as those initiated by pancreatic enzymes (Nouha et al., 2022). After absorption in the small intestine or colon, compounds are further metabolized at the enterocyte level. There are different reports where microencapsulation has been used with probiotics and prebiotics, which increase their storage stability within the human intestine, an important factor for reaching the colon (Yao et al., 2020). Subsequently, the compounds can be released at the action site (Braim et al., 2019). Some products reach the liver through the portal vein, where they undergo further metabolism before being transported back into the blood and eventually discarded in the urine (Grgić et al., 2020).

The microbiota and its interaction with other dietary compounds such as milk, carbohydrates, or phytic acid from fibers also play a fundamental role. This interaction causes bioactive compounds to undergo changes in their chemical structure, thereby altering their bioavailability (Teodora et al., 2019). Similarly, fermentation processes can negatively affect the bioavailability of bioactive compounds like phenolic compounds. However, the products of these reactions usually exhibit higher bioactivity than their precursors (Grgić et al., 2020).

Therefore, foods or food matrices that are consumed and contain a large number of bioactive compounds do not necessarily exhibit a significant change or effect on the body due to their low bioavailability. Several factors, including age, gender, physiological conditions, colonic microflora, and enzymatic activity, can influence bioavailability.

As a result, alternative approaches have been sought to increase bioavailability. One such approach that has gained significant relevance in recent years is encapsulation. Encapsulation has been shown to be an effective method for delivering treatments, as it enables controlled release of active ingredients at specific sites of interest within the gastrointestinal tract. Encapsulated bioactive compounds interact with water, oxidative stress, heat, strong acids, and bile, enhancing their effects (Liu et al., 2017; Ferrer-Sierra et al., 2022; Guan et al., 2015; Kusmayadi et al., 2019; Martin et al., 2014; Vlachopoulos et al., 2022).

This makes microencapsulated bioactive compounds a viable option for the treatment of diseases related to the colon or digestive tract (Gutierrez-Alvarado et al., 2022).

Currently, the search and formulation of foods or products with bioactive activities have been topics of interest in order to treat various health problems in different populations. One of the most important factors is the bioaccessibility of nutrients or compounds. To describe the potential activities, various *in vitro* and *in vivo* models have been described, which illustrate the potential effects that a food or compound can have (Ochoa Becerra et al., 2020).

5.5 ADVANTAGES OF COMPOUND ENCAPSULATION

The multiple effects that bioactive compounds can present offer an attractive alternative for the treatment of various ailments. However, limitations related to the stability of these molecules have led to research efforts aimed at producing stable products with notable effects. To achieve this, the search for methods of microencapsulation formulation allows for obtaining components with greater bioavailability and bioactivity compared to their untreated counterparts. This technique has successfully achieved the aforementioned objective by more than doubling bioavailability and bioaccessibility, resulting in enhanced effect and stability (Artiga-Artigas et al., 2018; Singh et al., 2018; Wang et al., 2020).

Furthermore, these types of formulations exhibit characteristics of prolonged release, making them an attractive option for the delivery of different drugs in long-term treatments in localized areas after stomach digestion, as well as for cancer treatment (de Oliveira Pedro et al., 2018; Wang et al., 2018). Microencapsulation has also been found to provide greater resistance to temperature, light, different pH levels, as well as improved bioavailability, bioaccessibility, chemical stability, and protection against evaporation. Additionally, it can bring about changes in organoleptic characteristics such as color, flavor, and smell, while increasing storage time and solubility (Figure 5.2) (Miranda et al., 2019; Chhikara et al., 2018; Chutia & Mahanta, 2021; de Oliveira Pedro et al., 2018; Mansour et al., 2020; Sri et al., 2013; Tarone et al., 2020; Wang et al., 2018; Yan et al., 2021).

Some of the conventional techniques for microencapsulation fail to fully provide protection and stability to encapsulated compounds. Therefore, adaptations are sought to solve this type of situation. Some of these adaptations include the formulation of nanoemulsions and nanoliposomes, which are nanosystems widely used in the pharmaceutical and food industries. These nanosystems not only provide better protection to the bioactive compound, but also increase their bioavailability (Chen & Inbaraj, 2019).

Furthermore, it has been shown that nanoemulsions and nanoliposomes can improve the release of highly unstable compounds under gastrointestinal conditions. They also provide high thermal stability, longer storage time, and overall stability. It is worth mentioning that the materials used as matrices or envelopes are usually inexpensive, making them a good option for incorporating these compounds into foods and functional additives (Ahmad et al., 2018; Cai et al., 2019; Chau Miranda et al., 2019; Mansour et al., 2020).

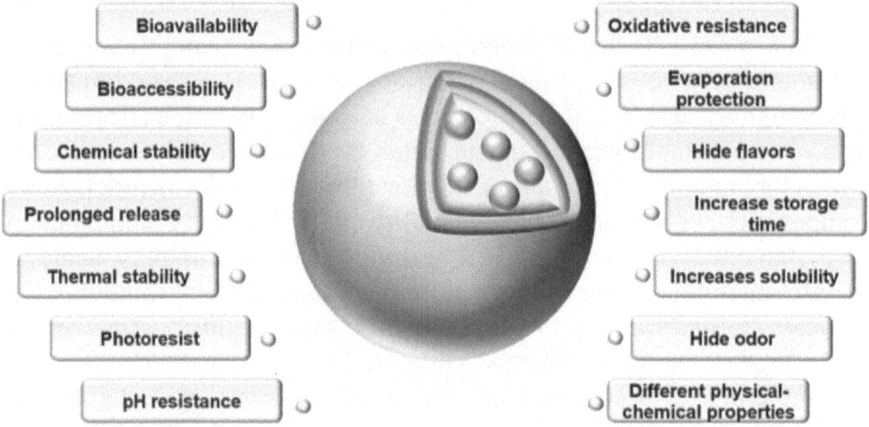

FIGURE 5.2 Advantages of microencapsulated compounds.

Source: Elaborated by the author.

5.6 APPLICATION OF MICROENCAPSULATED AS AN ANTI-INFLAMMATORY TREATMENT

Inflammation plays a fundamental role on pathogenesis of different diseases and is related to expression of different pro-inflammatory proteins which can be activated or deactivated depending on presence of different genetic or environmental factors (He et al., 2021). This condition can be related to different chronic diseases such as type II diabetes, cardiovascular diseases, lung diseases and cancer. Therefore, targeting inflammation is an attractive strategy to counteract the burden of health problems, which in many cases are related to dietary issues. Different studies consignee degrees of inflammation from mild to acute inflammation. In some cases, it is mentioned that, if this action cannot be treated or limited by the body, a prolonged condition occurs, the body presents a chronic pro-inflammatory condition, which is directly related to the activation of cytokines, and chemokines (IL-1β, IL-6, IL-10, and TNF-α), which in many cases are difficult to treat health problems, since it may be due to the presence of agents that promote inflammation as well as genetic issues (El-Abhar et al., 2008; He et al., 2021; Lee et al., 2017).

Currently, the application of microencapsulated compounds has been proposed to treat long-term problems or pathologies related to inflammation (Table 5.3). This approach offers a viable and attractive option for both the food and pharmaceutical industries. Microencapsulation allows for the release of active principles in localized areas, at different pH levels, and in the presence of diverse enzymatic activities. Furthermore, it enhances the effects of these compounds in the organism and improves their stability against various environmental stress factors (Miranda et al., 2019; Chhikara et al., 2018; Chutia & Mahanta, 2021; de Oliveira Pedro et al., 2018; Mansour et al., 2020; Sri et al., 2013; Tarone et al., 2020; Wang et al., 2018; Yan et al., 2021).

One notable function of encapsulating bioactive compounds is the treatment of various pathologies related to intestinal inflammation. This topic has gained significant relevance in recent years due to its increasing occurrence in society. The disease is characterized by localized inflammation in the intestine, overexpression of proinflammatory cytokines, and morphological changes at the cellular and molecular level (Carrasco-Avino, 2019; El-Abhar et al., 2008). By employing microencapsulation techniques, bioactive compounds can achieve greater stability and resistance to the digestive conditions that typically degrade them, thereby limiting or nullifying their anti-inflammatory effects (Grgić et al., 2020).

5.7 CONCLUSIONS

Microencapsulation represents a viable and novel option for the production of both food and pharmaceutical products. It is used when there is a need to contain or supply a bioactive compound with different properties and functions of interest in the organism. One of the highlighted properties of a bioactive compound is its anti-inflammatory activities. However, these activities are normally lost when the compound is supplied freely due to the gastrointestinal conditions it is subjected to. By incorporating the compound into a microcapsule, its effect can be potentiated. It can reach sites of action after digestion in the intestinal tract, altering the production of proinflammatory cytokines and regulating inflammation.

TABLE 5.3 Microencapsulation Work is Carried Out to Promote Anti-Inflammatory Activity

Matrix or Active Substance	Capsule Material	Technique	Effect	References
Bovine lactoferrin	Chitosan	Emulsification/ internal gelation	Reduction of cytotoxicity, delays the growth of parasites, reduction of deterioration in intestinal epithelial cells in humans and produces an anti-inflammatory effect.	Braim et al. (2019)
Echinacea purpurea extract	Chitosan-silica	Spray cooling	It has anti-inflammatory activities through the regulation of proinflammatory cytokines, reduction of iNOS production and immunomodulation.	Johnson et al. (2022)
Alcea rosea extract	Lectin	Liposomes	It regulates the expression of proinflammatory genes, as well as genes related to the antioxidant activity (COX2, iNOS, SOD, and GPx) and allows the increase of the population of microorganisms in the ileum.	Hassirian et al. (2022)
Alpha-lipoic acid (ALA)/ 1,2-distearoyl-sn-glycero-3-phosphoethanolamine-poly(ethylene glycol) (DSPE-PEG)	Polyamidoamine	Polymerization	It presents a controlled release of antioxidants under the stimulation of enzymes of pathological origin, it presents an enhanced antimicrobial effect and reduces the pH of the inflammatory microenvironment.	Wang et al. (2022)
Indirubin derivatives	Poly(D,L-lactide-co-glycolide) (PLGA)	Polymerization	It regulates the release of inflammation-relevant cytokines and lipid mediators from human monocytes.	Czapka et al. (2022)
Quercetin/curcumin/ hyaluronic acid/	Chitosan	Ionic gelation	It has a potential for tissue regeneration, superior efficacy in wound healing, as well as an anti-inflammation effect.	Hussain et al. (2022)

TABLE 5.3 *(Continued)*

Matrix or Active Substance	Capsule Material	Technique	Effect	References
Peach pomace extract	Whey protein, pea protein, maltodextrin, and gum Arabic.	Lyophilization	It presents an increase in antioxidant, antihyperglycemic and anti-inflammatory activities.	Vulic et al. (2022)
Extract of *Malpighia emarginata* and *Camellia sinensis* L.	Maltodextrin	Spray drying	Modulates the inflammatory response by altering the secretion of cytokines (IL-1β, IL-6, IL-10, and TNF-α), in addition, it has synergistic antioxidant and anti-inflammatory effects.	Cabral et al. (2020)
Quercetin	Zein	Precipitation	Increases the effectiveness of anti-inflammatory potential.	Penalva et al. (2016)
Resveratrol	Zein/lysine	Spray-drier	It presents an anti-inflammatory effect, as well as an increase in availability (19 times more), allowing for to reduction of endotoxic symptoms in murine models.	Penalva et al. (2015)
Quercetin	Chitosan	Ionic gelation	Significantly improves wound healing by modulating cytokines and growth factors involved in the inflammatory and proliferative phases of wound healing.	Choudhary et al. (2020)
C. carandas extract	Gelatin fiber	Electrospray	It exhibits better thermal resistance, higher drug release efficiency, as well as tyrosinase inhibitory activity.	Wiphadaa et al. (2021)
Sesamol	Stearic acid/chitosan	Lyophilization	It regulates the enzymatic and non-enzymatic antioxidant activity and modifies inflammatory signaling as well as the apoptotic pathway through the NF-κB signaling pathway.	Wang et al. (2021)

TABLE 5.3 *(Continued)*

Matrix or Active Substance	Capsule Material	Technique	Effect	References
Bitter melon juice	Maltodextrin/gum Arabic, gum Arabic/pectin, pectin/soy protein isolate.	Spray drying	It improves the antioxidant and antidiabetic activity, in addition, it significantly decreases the levels of IL-1β, IL-6, and TNF-α.	Wang et al. (2021)
Capsaicinoids from *Capsicum annuum*	Rice bran phospholipids.	Lyophilization	Reduces capsaicinoid irritation by half and improved anti-inflammatory activity 10 times more.	Abulencia et al. (2020)
Resveratrol	Zein/pectin	Precipitation	Improves anti-inflammatory activity by inhibiting the production of NO, PGE, IL-1β, IL-6, TNF-α, TLR4 and inhibiting phosphorylation of JNK, ERK1/2, p38 and MAPK, promotes the release of IL-10.	Liu et al. (2020)

ACKNOWLEDGMENTS

JDGO thanks the National Council of Science and Technology (CONACyT)) –Mexico for the financial support during their PhD in Food Science and Technology at UAdeC.

CONFLICT OF INTEREST

The authors declare no conflict of interest.

KEYWORDS

- **anti-inflammatory treatment**
- **bioactive compounds**
- **bioavailability**
- **formulation**
- **inflammatory conditions**
- **microencapsulation**
- **microparticles**
- **proinflammatory cytokines**

REFERENCES

Abdel-Aal, E. S. M., Young, J. C., & Rabalski, I., (2006). Anthocyanin composition in black, blue, pink, purple, and red cereal grains. *Journal of Agricultural and Food Chemistry*, *54*(13), 4696–4704. https://doi.org/10.1021/jf0606609.

Ahmad, M., Ashraf, B., Gani, A., & Gani, A., (2018). Microencapsulation of saffron anthocyanins using β glucan and β cyclodextrin: Microcapsule characterization, release behavior & antioxidant potential during in-vitro digestion. *International Journal of Biological Macromolecules*, *109*, 435–442. https://doi.org/10.1016/j.ijbiomac.2017.11.122.

Araba, M., MarziehHosseinib, S., Nayebzadehb, K., Khorshidian, N., Yousefic, M., Seyed, H. R., & Mortazavianb, A. M., (2019). Microencapsulation of microbial canthaxanthin with alginate and high methoxyl pectin and evaluation the release properties in neutral and acidic condition. *International Journal of Biological Macromolecules, 691–698.*

Artiga-Artigas, M., Odriozola-Serrano, I., Oms-Oliu, G., & Martín-Belloso, O., (2018). Nano-structured systems to increase bioavailability of food ingredients. In *Nanomaterials for Food Applications*. Elsevier Inc. https://doi.org/10.1016/B978-0-12- 814130-4.00002-6.

Bobek, P., Nosál'Ová, V., & Černa, S., (2001). Influence of diet containing extract of black elder (*Sambucus nigra*) on colitis in rats. *Biologia, 56*(6), 643–648.

Bowen, K. J., Kris-Etherton, P. M., West, S. G., Fleming, J. A., Connelly, P. W., Lamarche, B., Couture, P., et al., (2019). Diets enriched with conventional or high-oleic acid canola oils lower atherogenic lipids and lipoproteins compared to a diet with a western fatty acid profile in adults with central adiposity. *The Journal of Nutrition, Nutrition and Disease, 143*(3), 471–478. https://doi.org/10.1093 / jn / nxy307.

Braim, S., Śpiewak, K., Brindell, M., Heeg, D., Alexander, C., & Monaghan, T., (2019). Lactoferrin-loaded alginate microparticles to target clostridioides difficile infection. *Journal of Pharmaceutical Sciences, 108*(7), 2438–2446. https://doi.org/10.1016/J.XPHS.2019.02.025.

Cabral, N., Natan, E., Nascimento, D. O., Bezerra, I., Oliveira, D., Miguel, H., Oliveira, L., et al., (2020). Anti-inflammatory and antixidant properties of blend formulated with compounds of *Malpighia emarginata* D. C (acerola) and *Camellia sinensis* L. (green tea) in lipopolysaccharide-stimulated RAW 264. 7 macrophages. *Biomedicine & Pharmacotherapy, 128*, 110277. https://doi.org/10.1016/j.biopha.2020.110277.

Cai, X., Du, X., Cui, D., Wang, X., Yang, Z., & Zhu, G., (2019). Improvement of stability of blueberry anthocyanins by carboxymethyl starch/xanthan gum combinations microencapsulation. *Food Hydrocolloids, 91*, 238–245. https://doi.org/10.1016/j.foodhyd.2019.01.034.

Carrasco-Avino, G., (2019). Histología en la enfermedad inflamatoria intestinal. *Revista Médica Clínica Las Condes, 30*(4), 283–298. https://doi.org/10.1016/j.rmclc.2019.06.005.

Castro-Rosas, J., Ferreira-Grosso, C. R., Gómez-Aldapa, C. A., Rangel-Vargas, E., Rodríguez-Marín, M. L., Guzmán-Ortiz, F. A., & Falfan-Cortes, R. N., (2017). Recent advances in microencapsulation of natural sources of antimicrobial compounds used in food: A review. *Food Research International*, 575–587. https://doi.org/10.1016/j.foodres.2017.09.054.

Chau, M. G., Herrera, C. O., & Condorhuamán, F. M., (2019). Actividad antioxidante in vitro, de diferentes extractos del fruto de *Physalis peruviana* L. (aguaymanto). *Revista Peruana de Medicina Integrativa, 4*(1), 22. https://doi.org/10.26722/rpmi.2019.41.105.

Chen, B. H., & Inbaraj, B. S., (2019). Nanoemulsion and nanoliposome based strategies for improving anthocyanin stability and bioavailability. *Nutrients, 11*(5). https://doi.org/10.3390/nu11051052.

Chhikara, N., Kaur, R., Jaglan, S., Sharma, P., Gat, Y., & Panghal, A., (2018). Bioactive compounds and pharmacological and food applications of: *Syzygium cumini*: A review. *Food and Function, 9*(12), 6096–6115. https://doi.org/10.1039/c8fo00654g.

Choudhary, A., Kant, V., Lal, B., & Joshi, V. G., (2020). Quercetin loaded chitosan tripolyphosphate nanoparticles accelerated cutaneous wound healing in Wistar rats. *European Journal of Pharmacology, 880*, 173172. https://doi.org/10.1016/j.ejphar.2020.173172.

Choudhury, N., Meghwal, M., & Das, K., (2021). Microencapsulation: An overview on concepts, methods, properties, and applications in foods. *Food Frontiers, 2*, 426–442. https://doi.org/10.1002/fft2.94.

Chutia, H., & Mahanta, C. L., (2021). Properties of starch nanoparticle obtained by ultrasonication and high pressure homogenization for developing carotenoids-enriched powder and Pickering nanoemulsion. *Innovative Food Science and Emerging Technologies, 74,* 102822. https://doi.org/10.1016/j.ifset.2021.102822.

Chuyen, H. V., Roach, P. D., Golding, J. B., Parks, S. E., & Nguyen, M. H., (2019). Encapsulation of carotenoid-rich oil from Gac peel: Optimization of the encapsulating process using a spray drier and the storage stability of encapsulated powder. *Powder Technology, 344,* 373–379. https://doi.org/10.1016/j.powtec.2018.12.012.

Czapka, A., Grune, C., Schädel, P., Bachmann, V., Scheuer, K., Dirauf, M., Weber, C., et al., (2022). Drug delivery of 6-bromoindirubin-3 '-glycerol-oxime ether employing poly (d, l-lactide-co-glycolide)-based nanoencapsulation techniques with sustainable solvents. *Journal of Nanobiotechnology, 7,* 1–21. https://doi.org/10.1186/s12951-021-01179-7.

De Oliveira, P. R., Hoffmann, S., Pereira, S., Goycoolea, F. M., Schmitt, C. C., & Neumann, M. G., (2018). Self-assembled amphiphilic chitosan nanoparticles for quercetin delivery to breast cancer cells. *European Journal of Pharmaceutics and Biopharmaceutics,* 203–210. https://doi.org/10.1016/J.EJPB.2018.08.009.

El-Abhar, H. S., Hammad, L. N. A., & Gawad, H. S. A., (2008). Modulating effect of ginger extract on rats with ulcerative colitis. *Journal of Ethnopharmacology, 118*(3), 367–372. https://doi.org/10.1016/j.jep.2008.04.026.

El-Baz, F. K., Aly, H. F., & Abd-Alla, H. I., (2020). The ameliorating effect of carotenoid rich fraction extracted from *Dunaliella salina* microalga against inflammation-associated cardiac dysfunction in obese rats. *Toxicology Reports, 7,* 118–124. https://doi.org/10.1016/j.toxrep.2019.12.008.

Favaro-trindade, C. S., De Matos, Jr. F. E., Okuro, P. K., Dias-ferreira, J., Cano, A., Severino, P., Zielinska, A., & Souto, E. B., (2021). Encapsulation of active pharmaceutical ingredients in lipid micro / nanoparticles for oral administration by spray-cooling. *Pharmaceutics, 13,* 1186. https://doi.org/10.3390/pharmaceutics13081186.

Fernanda, T. M., Pavez, C., & Parada, A., (2020). Microbiota, dietary habits, and diet in inflammatory bowel disease. *Revista Chilena de Nutricion, 47*(5), 822–829. https://doi.org/10.4067/S0717-75182020000500822.

Ferrer-Sierra, M., Rodriguez-Lopez, P., Leyva-Jimenez, F. J., Borras-Linares, I., Giacomazza, D., Fredes, C., Robert, C. P. S., et al., (2022). Encapsulation technologies applied to bioactive phenolic compounds and probiotics with potential application on chronic inflammation. *Current Advances for Development of Functional Foods Modulating Inflammation and Oxidative Stress,* 447–476. https://doi.org/10.1016/B978-0-12-823482-2.00010-8.

Gañan, N., Gabriela, M., Ribotta, P. D., & González, A., (2020). Study of chia oil microencapsulation in soy protein microparticles using supercritical CO_2-assisted impregnation. *Journal of CO_2 Utilization, 40,* 101221. https://doi.org/10.1016/j.jcou.2020.101221.

Gonzalez-Reza, R. M., Garcia-Betanzos, C. I., Sánchez-Valdes, L. I., Quintanar-Guerrero, D., Cornejo-Villegas, M. A., & Zambrano-Zaragoza, M. L., (2018). The functionalization of nanostructures and their potential applications in edible coatings. *Coatings, 8,* 160. https://doi.org/10.3390/coatings8050160.

Grgić, J., Šelo, G., Planinić, M., Tišma, M., & Bucić-Kojić, A., (2020). Role of the encapsulation in bioavailability of phenolic compounds. *Antioxidants, 9*(10), 1–36. https://doi.org/10.3390/antiox9100923.

Guan, Q., Chen, W., & Hu, X., (2015). Development of lovastatin-loaded poly(lactic acid) microspheres for sustained oral delivery: In vitro and ex vivo evaluation. *Drug Design, Development and Therapy, 9,* 791–798. https://doi.org/10.2147/DDDT.S76676.

Guo, F., Tsao, R., Li, C., Wang, X., Zhang, H., Jiang, L., Sun, Y., & Xiong, H., (2022). Polyphenol content of green pea (*Pisum sativum* L.) hull under in vitro digestion and effects of digestive products on anti-inflammatory activity and intestinal barrier in the caco-2/Raw264.7 coculture model. *Journal of Agricultural and Food Chemistry*, *70*(11), 3477–3488. https://doi.org/10.1021/acs.jafc.2c00102.

Gutierrez-Alvarado, K., Chacón-Cerdas, R., & Starbird-Perez, R., (2022). Pectin microspheres: Synthesis methods, properties, and their multidisciplinary applications. *Chemistry*, *4*, 121–136. https://doi.org/10.3390/chemistry40100.

Hassirian, N., Karimi, E., & Oskoueian, E., (2022). Nanoliposome-encapsulated phenolic-rich fraction from *Alcea rosea* as a dietary phytobiotic in mice challenged by *Escherichia coli*. *Annals of Microbiology*, *72*(6), 1–11. https://doi.org/10.1186/s13213-022-01665-9.

He, Z., Chen, L., Catalan-Dibene, J., Bongers, G., Faith, J. J., Suebsuwong, C., DeVita, R. J., et al., (2021). Food colorants metabolized by commensal bacteria promote colitis in mice with dysregulated expression of interleukin-23. *Cell Metabolism*, *33*(7), 1358–1371.e5. https://doi.org/10.1016/j.cmet.2021.04.015.

Hussain, Z., Pandey, M., Ei, H., Kaur, T., Wei, G., Cheau, P., Mei, T., & Abourehab, M. A. S., (2022). Hyaluronic acid functionalization improves dermal targeting of polymeric nanoparticles for management of burn wounds: *In vitro*, ex vivo and in vivo evaluations. *Biomedicine & Pharmacotherapy*, *150*, 112992. https://doi.org/10.1016/j.biopha.2022.112992.

Johnson, A., Huang, Y. C., Mao, C. F., Chen, C. K., & Thomas, S., (2022). Protective effect of ethanolic extract of *Echinacea purpurea* contained nanoparticles on meniscal / ligamentous injury induced osteoarthritis in obese male rats. *Scientific Reports*, *0123456789*, 1–15. https://doi.org/10.1038/s41598-022-09380-w.

Kashappa, G. D. H., & Hyun, J. P., (2005). Recent developments in microencapsulation of food ingredients. *Drying Technology*, *23*(7), 1361–1394. https://doi.org/10.1081/DRT-200063478.

Kaur, P., Elsayed, A., Subramanian, J., & Singh, A., (2021). Encapsulation of carotenoids with sucrose by co-crystallization: Physicochemical properties, characterization, and thermal stability of pigments. *LWT*, *140*, 110810. https://doi.org/10.1016/j.lwt.2020.110810.

Konstantinidi, M., & Koutelidakis, A. E., (2019). Functional foods and bioactive compounds: A review of its possible role on weight management and obesity's metabolic consequences. *Medicines*, *6*(3), 94. https://doi.org/10.3390/medicines6030094.

Kusmayadi, A., Adriani, L., Abun, A., Muchtaridi, M., & Tanuwiria, U. H., (2019). The microencapsulation of mangosteen peel extract with maltodextrin from Arenga starch: Formulation and characterization. *Journal of Applied Pharmaceutical Science*, *9*(3), 33–40. https://doi.org/10.7324/JAPS.2019.90306.

Lee, Y. M., Yoon, Y., Yoon, H., Park, H. M., Song, S., & Yeum, K. J., (2017). Dietary anthocyanins against obesity and inflammation. *Nutrients*, *9*(10), 1–15. https://doi.org/10.3390/nu9101089.

Liu, H., Cui, S. W., Chen, M., Liang, R., & Xu, F., (2017). Protective approaches and mechanisms of microencapsulation to the survival of probiotic bacteria during processing, storage and gastrointestinal digestion: A review. *Critical Reviews in Food Science and Nutrition*, *59*(17), 2863–2878. https://doi.org/10.1080/10408398.2017.1377684.

Liu, Y., Liang, X., Zou, Y., Peng, Y., Mcclements, D. J., & Hu, K., (2020). Resveratrol-loaded biopolymer core–shell nanoparticles: Bioavailability and anti-inflammatory effects. *Food & Function, 11*, 4014–4025. https://doi.org/10.1039/d0fo00195c.

Mansour, M., Salah, M., & Xu, X., (2020). Effect of microencapsulation using soy protein isolate and gum arabic as wall material on red raspberry anthocyanin stability, characterization, and simulated gastrointestinal conditions. *Ultrasonics Sonochemistry, 63*, 104927. https://doi.org/10.1016/j.ultsonch.2019.104927.

Maroulas, P., Kripotou, S., Pissis, P., Fainleib, A., Bei, I., Bershtein, V., & Gomza, Y., (2009). Molecular mobility in polycyanurate/clay nanocomposites studied by dielectric techniques. *Journal of Composite Materials, 43*(9), 943–958. https://doi.org/10.1177/0021998308097736.

Martin, M. J., Federico, L. V., Ruiz, M. A., & Morales, M. E., (2014). Microencapsulation of bacteria: A review of different technologies and their impact on the probiotic effects. *Innovative Food Science and Emerging Technologies*. https://doi.org/10.1016/j.ifset.2014.09.010.

Mohammed, N. K., Tan, C. P., Manap, Y. A., Muhialdin, B. J., Shobirin, A., & Hussin, M., (2020). Spray drying for the encapsulation of oils — A review. *Molecules, 25*, 3873. https://doi.org/10.3390/molecules25173873.

Nahum, V., & Domb, A. J., (2021). Recent developments in solid lipid microparticles for food ingredients delivery. *Foods, 10*, 400. https://doi.org/10.3390/foods10020400.

Nesterenko, A., Alric, I., Silvestre, F., & Durrieu, V., (2013). Vegetable proteins in microencapsulation: A review of recent interventions and their effectiveness. *Industrial Crops and Products, 42*(1), 469–479. https://doi.org/10.1016/j.indcrop.2012.06.035.

Niño-Vásquez, I. A., Muñiz-Márquez, D., Ascacio-Valdés, J. A., Contreras-Esquivel, C., Aguilar, C. N., Rodríguez-Herrera, R., Flores-Gallegos, A. C., et al., (2022). Co-microencapsulation: A promising multi-approach technique for enhancement of functional properties. *Bioengineered, 13*(3), 5168–5189. https://doi.org/10.1080/21655979.2022.2037363.

Nouha, D., Meryam, C., Khaled, A., Sabrine, C., Sana, G., Mouna, T., Tahia, B., Habib, A., & Hanen, B.-K. (2022). β-cyclodextrin microencapsulation enhanced antioxidant and antihyperlipidemic properties of Tunisian *Periploca angustifolia* roots condensed tannins in rats. *Environmental Science and Pollution Research*. https://doi.org/10.1007/s11356-022-20095-3.

Oancea, A. M., Hasan, M., Vasile, A. M., Barbu, V., Enachi, E., Bahrim, G., Râpeanu, G., et al., (2018). Functional evaluation of microencapsulated anthocyanins from sour cherries skins extract in whey proteins isolate. *LWT, 95*, 129–134. https://doi.org/10.1016/j.lwt.2018.04.083.

Ochoa, B. M., Mojica, C. L., Hsieh, L. M., Mateos, D. J., & Castillo, H. G., (2020). Lutein as a functional food ingredient: Stability and bioavailability. *Journal of Functional Foods, 66*, 103771. https://doi.org/10.1016/j.jff.2019.103771.

Otálora, M. C., Wilches-torres, A., & Gómez, C. J. A., (2022). Spray-drying microencapsulation of pink guava (*Psidium guajava*) carotenoids using mucilage from opuntia *Ficus-indica* cladodes and aloe vera leaves as encapsulating materials. *Polymers, 14*, 310. https://doi.org/10.3390/polym14020310.

Ozkan, G., Franco, P., De Marco, I., Xiao, J., & Capanoglu, E., (2019). A review of microencapsulation methods for food antioxidants: Principles, advantages, drawbacks, and applications. *Food Chemistry*, *272*, 494–506. https://doi.org/10.1016/j.foodchem.2018.07.205.

Paulo, F., & Santos, L., (2017). Design of experiments for microencapsulation applications: A review. *Materials Science and Engineering C*, *77*, 1327–1340. https://doi.org/10.1016/j.msec.2017.03.219.

Penalva, R., Esparza, I., Larraneta, E., Gamazo, C., & Irache, J. M., (2015). Zein-based nanoparticles improve the oral bioavailability of resveratrol and its anti-inflammatory effects in a mouse model of endotoxic shock. *Journal of Agricultural and Food Chemistry*, *63*(23), 5603–5611. https://doi.org/10.1021/jf505694e.

Penalva, R., González-navarro, C. J., Gamazo, C., Esparza, I., & Irache, J. M., (2016). Zein nanoparticles for oral delivery of quercetin: Pharmacokinetic studies and preventive anti-inflammatory effects in a mouse model of endotoxemia. *Nanomedicine: Nanotechnology, Biology, and Medicine*, *13*, 103–110. https://doi.org/10.1016/j.nano.2016.08.033.

Revuelta, M. V., Bogdan, S., Gámez-Espinosa, E., Deyá, M. C., & Romagnoli, R., (2021). Green antifungal waterborne coating based on essential oil microcapsules. *Progress in Organic Coatings*, *151*(2). https://doi.org/10.1016/j.porgcoat.2020.106101.

Ribeiro, J. S., & Veloso, C. M., (2021). Microencapsulation of natural dyes with biopolymers for application in food: A review. In: *Food Hydrocolloids* (Vol. 112). Elsevier B.V. https://doi.org/10.1016/j.foodhyd.2020.106374.

Richards, K., & Malik, D. J., (2021). Microencapsulation of bacteriophages using membrane emulsification in different pH-triggered controlled release formulations for oral administration. *Pharmaceuticals*, *14*, 424. https://doi.org/10.3390/ph14050424.

Rodrigues, R., Bilibio, D., Plata-oviedo, M. S. V., Pereira, E. A., Mitterer-dalto, M. L., Perin, E. C., & Carpes, S. T., (2021). Microencapsulated and lyophilized propolis co-product extract as antioxidant synthetic replacer on traditional Brazilian starch biscuit. *Molecules*, *26*, 6400. https://doi.org/10.3390/molecules26216400.

Rodríguez, G. T., Peña, G. M., Gómez, T. N., Santisteban, L. Y., & Hernández, T. M., (2015). Estrés oxidativo: Genética, dieta y desarrollo de enfermedades. *Correo Científico Médico de Holguín*, *19*(4), 1–16.

Sadaji, D. S., Alemzadeh, I., Sadar, V. A., & Vossoughi, A., (2020). Optimization of alginate-whey protein isolate microcapsules for survivability and release behavior of probiotic bacteria. *Applied Biochemistry and Biotechnology*, *190*, 182–196. https://doi.org/10.1007/s12010-019-03071-5.

Sepúlveda-Vázquez, J., Torres-Acosta, J. F., Sandoval-Castro, C. A., Martínez-Puc, J. F., & Chan-Pérez, J. I., (2018). La importancia de los metabolitos secundarios en el control de nematodos gastrointestinales en ovinos con énfasis en Yucatán, México. *Journal of the Selva Andina Animal Science*, *5*(2), 79–95. https://doi.org/10.36610/j.jsaas.2018.050200079.

Sharif, N., Khoshnoudi-Nia, S., & Jafari, S. M., (2020). Nano/microencapsulation of anthocyanins; a systematic review and meta-analysis. *Food Research International*, *132*, 109077. https://doi.org/10.1016/j.foodres.2020.109077.

Sharma, M., Hussain, S., Shalima, T., Aav, R., & Bhat, R., (2022). Valorization of seabuckthorn pomace to obtain bioactive carotenoids: An innovative approach of using green extraction techniques (ultrasonic and microwave-assisted extractions) synergized with green solvents (edible oils). *Industrial Crops and Products*, *175*, 114257. https://doi.org/10.1016/j.indcrop.2021.114257.

Singh, H., Kumar, C., Singh, N., Paul, S., & Jain, S. K., (2018). Nanoencapsulation of docosahexanoic acid (DHA) using combination of food grade polymeric wall materials and its application for improvement in bioavailability and oxidative stability. *Food and Function, 9*(4), 2213–2227. https://doi.org/10.1039/C7FO01391D.

Sri, S, J., Seethadevi, A., Suria, P. K., Muthuprasanna, P., & Pavitra, P., (2013). Microencapsulation: A review. *Research Journal of Pharmacy and Technology, 6*(9), 954–961. https://ijpbs.net/abstract.php?article=MTI2MA== (accessed on 24 February 2024).

Stefanescu, B. E., Nemes, S. A., Teleky, B. E., Calinoiu, L. F., Mitrea, L., Adrian, G. A., Szabo, K., et al., (2022). Microencapsulation and bioaccessibility of phenolic compounds of vaccinium leaf extracts. *Antioxidants, 11*, 674. https://doi.org/10.3390/antiox11040674.

Suganya, V., & Anuradha, V., (2017). Microencapsulation and nanoencapsulation: A review. *International Journal of Pharmaceutical and Clinical Research, 9*(3), 233–239. https://doi.org/10.25258/ijpcr.v9i3.8324.

Tarone, A. G., Cazarin, C. B. B., & Marostica, Jr. M. R., (2020). Anthocyanins: New techniques and challenges in microencapsulation. *Food Research International, 133*. https://doi.org/10.1016/j.foodres.2020.109092.

Tena, N., Martín, J., & Asuero, A. G., (2020). State of the art of anthocyanins: Antioxidant activity, sources, bioavailability, and therapeutic effect in human health. *Antioxidants, 9*(5). https://doi.org/10.3390/antiox9050451.

Teodora, C., Péter, N., Ganea, C., János, S., & Mocanu, M. M., (2019). Molecular mechanisms and bioavailability of polyphenols in prostate cancer. *Molecular Sciences, 20*, 1062. https://doi.org/10.3390/ijms20051062.

Trang, N. T. T., Anh Le, T. V., Dang, N. N., Chi Nguyen, D., Nhan, N. P. T., Truc, T. T., Vinh, N. Q., et al., (2021). Microencapsulation of essential oils by spray-drying and influencing factors. *Journal of Food Quality, 2021*, 15. https://doi.org/10.1155/2021/5525879 Review.

Vlachopoulos, A., Karlioti, G., Balla, E., Daniilidis, V., Kalamas, T., Stefanidou, M., Bikiaris, N. D., et al., (2022). Poly(lactic acid)-based microparticles for drug delivery applications: An overview of recent advances. *Pharmaceutics, 14*(2), 1–37. https://doi.org/10.3390/pharmaceutics14020359.

Von, L. J., Moon, J., Lee, J., & Ramkumar, S., (2020). Carotenoid metabolism at the intestinal barrier. *Biochimica et Biophysica Acta – Molecular and Cell Biology of Lipids, 1865*(11), 158580. https://doi.org/10.1016/j.bbalip.2019.158580.

Vulic, J., Bibovski, K., Vanja, S., Kovacevic, S., Banjac, M. K., Canadovic-Brunet, J., Cetkovic, G., et al., (2022). Chemical and biological properties of peach pomace. *Processes, 10*, 642. https://doi.org/10.3390/pr10040642.

Wang, C., Chen, X., Nakamura, Y., Yu, C., & Qi, H., (2020). Fucoxanthin activities motivate its nano/micro-encapsulation for food or nutraceutical application: A review. *Food and Function, 11*(11), 9338–9358. https://doi.org/10.1039/d0fo02176h.

Wang, L., Clardy, A., Hui, D., & Wu, Y., (2021). Physiochemical properties of encapsulated bitter melon juice using spray drying. *Bioactive Carbohydrates and Dietary Fiber, 26*, 100278. https://doi.org/10.1016/j.bcdf.2021.100278.

Wang, L., Li, Y., Ren, M., Wang, X., Li, L., Liu, F., Lan, Y., et al., (2022). Bioactive materials pH and lipase-responsive nanocarrier-mediated dual drug delivery system to treat periodontitis in diabetic rats. *Bioactive Materials, 18*(1), 254–266. https://doi.org/10.1016/j.bioactmat.2022.02.008.

Wang, N., Yu, H., Song, Q., Mao, P., Li, K., & Bao, G., (2021). Sesamol-loaded stearic acid-chitosan nanomicelles mitigate the oxidative stress-stimulated apoptosis and induction of pro-inflammatory cytokines in motor neuronal of the spinal cord through NF-κB signaling pathway. *International Journal of Biological Macromolecules, 186*(277), 23–32. https://doi.org/10.1016/j.ijbiomac.2021.06.171.

Wang, Y., Wang, Y., Wang, J., Lei, W., Li, K., Wu, D., & Wang, X., (2018). Pharmacokinetics, biodistribution, and bioavailability of gossypol-loaded Pluronic® F127 nanoparticles. *Journal of Drug Delivery Science and Technology, 45*, 388–396. https://doi.org/10.1016/j.jddst.2018.04.002.

Wiphadaa, R., Nakorna, N., Choladdaa, S., Nophawanb, P., & Withawat, M., (2021). Electrospun gelatin fiber mats mixed with *C. carandas* extract and its enhanced stability and bioactivity. *Journal of Pharmaceutical Sciences, 110*(6), 2405–2415. https://doi.org/10.1016/j.xphs.2020.12.031.

World Health Organization (WHO), (2021). *The Top 10 Causes of Death.* https://www.who.int/es/news-room/fact-sheets/detail/the-top-10-causes-of-death (accessed on 24 February 2024).

Yan, C., Wu, X., Wang, Y., Peng, S., Chen, J., Zou, L., McClements, D. J., & Liu, W., (2021). Utilization of polysaccharide-based high internal phase emulsion for nutraceutical encapsulation: Enhancement of carotenoid loading capacity and stability. *Journal of Functional Foods, 84*, 104601. https://doi.org/10.1016/j.jff.2021.104601.

Yang, Y. H., Teresa, M., & López, I., (2021). *Dieta Mediterránea y dieta japonesa vs enfermedades neurodegenerativas Dieta Mediterránea y dieta japonesa vs enfermedades neurodegenerativas Mediterranean and Japanese diets vs. Neurodegenerative Diseases*, 1110–1148. https://doi.org/10.19230/jonnpr.3934.

Yao, M., Xie, J., Du, H., McClements, D. J., Xiao, H., & Li, L., (2020). Progress in microencapsulation of probiotics: A review. In: *Comprehensive Reviews in Food Science and Food Safety* (Vol. 19, No. 2, pp. 857–874). Blackwell Publishing Inc. https://doi.org/10.1111/1541-4337.12532.

Zhang, J., Wen, C., Zhang, H., Duan, Y., & Ma, H., (2020). Recent advances in the extraction of bioactive compounds with subcritical water: A review. *Trends in Food Science and Technology, 95*, 183–195. https://doi.org/10.1016/j.tifs.2019.11.018.

Zhang, R., Zhou, L., Li, J., Oliveira, H., Yang, N., Jin, W., Zhu, Z., Li, S., & He, J., (2020). LWT-food science and technology microencapsulation of anthocyanins extracted from grape skin by emulsification/internal gelation followed by spray/freeze-drying techniques: Characterization, stability and bioaccessibility. *LWT–Food Science and Technology, 123*, 109097. https://doi.org/10.1016/j.lwt.2020.109097.

CHAPTER 6

Sustainable Use of Andean Tubers: Nutritional, Bioactive, and Functional Properties

STEFFANNY SANCHEZ-PORTILLO,[1]
MARGARITA DEL ROSARIO SALAZAR-SÁNCHEZ,[2]
JUAN ALBERTO ASCACIO-VALDÉS,[1]
JOSÉ FERNANDO SOLANILLA-DUQUE,[3]
ADRIANA CAROLINA FLORES-GALLEGOS,[1] and
RAÚL RODRÍGUEZ-HERRERA[1]

[1]*Facultad de Ciencias Químicas, Universidad Autónoma de Coahuila, Unidad Saltillo, Saltillo, Coahuila, México*

[2]*Programa de Ingeniería Agroindustrial, Facultad de Ingenierías Universidad del Popular del Cesar, Cesar, Colombia*

[3]*Departamento de Agroindustrial, Facultad de Ciencias Agrarias, Universidad del Cauca, Cauca, Colombia*

ABSTRACT

Andean tubers are commonly recognized as edible underground roots or stems cultivated in the Andean region of the Americas. These crops are an important source of nutrients and bioactive compounds and have been used throughout history by the native people of this region. However, few are recognized outside of this geographical area. Recently, there has been increased interest in their chemical composition, as well as their effects and potential health benefits for consumers. This review focuses on detailing

Sustainable Environment and Health: Practical Strategies. Leonardo Sepúlveda Torre, Ann Rose Abraham, Neha Kanwar Rawat, & A. K. Haghi (Eds.)

the nutritional components, bioactive compounds, functional properties, and future consumption trends of five Andean tuber species (Arracacha, Mashua, Yacón, Oca, and Ulluco) of major commercial interest. The marked difference between one species and another in their nutritional and bioactive properties makes these tubers an excellent ingredient for the development of functional foods. In addition, the techno-functional properties of their components (e.g., starch) are relevant for the production of edible coatings and films. Their use outside the Andean region has been found to be very limited. For this reason, there is a need to implement strategies to increase the interest of other markets for their consumption. These tubers are a great opportunity for new research because of their techno-functional and bioactive properties.

6.1 INTRODUCTION

Tubers and roots are plant organs that contain parenchymatous tissue, where cells store water and carbohydrates in the form of starch. This makes them a valuable source of nutrients for both animals and humans. Tubers are classified as underground stems in botanical terms (Bonete & Urquizo, 2016). While the potato (*Solanum tuberosum*) is the most well-known and widely cultivated Andean tuber worldwide, this region is actually the center of origin for various tubers that differ in shape, color, and size. Unfortunately, these tubers remain relatively unknown outside of this region (Velásquez-Barreto et al., 2021). Despite being categorized as marginal crops (Clavijo Ponce, 2018), these Andean tubers have persisted over time. They are known for their resilience, being able to withstand drought, frost, and saline conditions. Cultivated at altitudes above 2,000 meters above sea level, they play a significant role in global food security (Solano et al., 2021).

There is a great variety of Andean tubers, some species are: achira (*Canna edulis*), ahipa (*Pachyrhizus ahipa*), arracacha (*Arracacia xanthorrhiza*), maca (*Lepidium meyenii*), mashua or isaño (*Tropaeolum tuberosum*), miso or mauca (*Mirabilis expansa*), oca (*Oxalis tuberosa*), ulluco, melloco or smooth potato (*Ullucus tuberosus*) and jicama or yacón (*Smallanthus sonchifolius*), which are distributed throughout most of South America (Velásquez-Barreto et al., 2018), Andean tubers have long been considered traditional crops. They form a significant part of the staple diet for the native people of this region of the American continent.

Primarily cultivated for self-consumption, they are also sold in smaller quantities (Parra-Quijano et al., 2012). These tubers are commonly consumed in soups, salads, or simply cooked (Velásquez-Barreto et al., 2021). They are known to be rich in nutrients such as carbohydrates, carotenoids, anthocyanins, and phenolic acids, with starch being the most abundant compound (Clavijo Ponce, 2018). Furthermore, they contain various chemical compounds that have the potential to contribute to the treatment of diseases and are considered as possible sources for obtaining chemical products in the pharmaceutical industry (Investigación & Citse, 2016). Fructo-oligosaccharides (FOS), identified in these plants, exhibit prebiotic, immunomodulatory, hypoglycemic, and hypolipidemic activities (Fabersani et al., 2018). Additionally, they are considered rich in inulin (a polysaccharide) (Aziz et al., 2021). Moreover, various phytochemical compounds such as anthocyanins, carotenoids, and phenols have been extracted from these crops (Pacheco et al., 2019). In addition, considering the increasing consumer demand for products or foods with higher nutritional value that contribute to health and well-being, and preferably derived from natural raw materials (Salazar et al., 2021), certain foods produced in the Andean region, such as tubers, rhizomes, and legumes (Heil et al., 2017), could serve as a source for the development of the food industry. These foods not only meet nutritional requirements but also promote their utilization, application, and exploitation (Salazar et al., 2021). This would help alleviate post-harvest losses experienced by producers due to the demands of a competitive market, which often leads to the wastage of these tubers due to factors such as size, shape, or color (Campos et al., 2018a).Therefore, the main objective of this chapter is to provide an overview of the nutritional and functional potential of certain Andean tubers, highlighting the benefits they offer to human health through the presence of phytochemicals. Additionally, their potential use in disease treatment will be explored, which will expand production alternatives, application, and marketing opportunities (Galindez et al., 2019). Ultimately, this will help reduce losses for producers and promote scientific advancements in the utilization of these species.

6.2 ANDEAN TUBERS

The Andean region of the American continent produces a wide variety of foods that were established by the indigenous people, some of which are the well-known Andean tubers (Bonete & Urquizo, 2016). These tubers were domesticated approximately 8,000 years ago in different parts of South America and present different compounds that have both medical and technological potential, which have not yet been fully exploited (Pacheco et al., 2020). They were grown at altitudes between 1,500 and 4,000 masl (Cruz-Tirado et al., 2019; Solano et al., 2021). In their structure, these tubers have fructooligosaccharides (Fabersani et al., 2018). Due to their content of various carotenoids, the Andean tubers have been used in anti-inflammatory processes (Heil et al., 2017), and for their glucosinates (Campos et al., 2018), these tubers have antioxidant capacity (Duque et al., 2017).

The nutritional value of Andean tubers varies depending on the variety, location, type of soil, and the agricultural management during crop production (Andrade-Mahecha et al., 2012). Although their main components are carbohydrates, some macro- and micro-nutrients have been reported. There are reports of some vitamins in their composition, which has boosted their consumption in the communities of the region, whether fresh or processed in soups, stews, and desserts (Velásquez-Barreto et al., 2021).

Among these Andean crops, the ibia (*Oxalis tuberosa* Mol.) (Morillo et al., 2019), *Smallantus sonchifolius*, *Ullucus tuberosus*, *Tropaeolum tuberosum*, and *Arracacia xanthorrhiza*, which in addition to its potential for technological, food, and industrial applications, advantages of this include resistance to drought and pests, low production costs due to minimal inputs required, and the possibility of establishing alternative crops such as legumes, which contribute to protection against diseases for these plants (Pacheco et al., 2020) (Figure 6.1).

6.2.1 GENERAL CHARACTERISTICS OF ANDEAN TUBERS

6.2.1.1 ARRACACIA XANTHORRHIZA

Arracacia xanthorrhiza is cultivated in different countries of South America (Brazil, Colombia, Ecuador, Peru, and Bolivia) (Chaali et al., 2020), at altitudes ranging from 600 to 3,900 masl, is commonly known as arracacha, celery, Peruvian parsnip, and potato analog (Quijano-Célis et al., 2016) is a staple food in the diets of Amazonian, Venezuelan Caribbean coast, and indigenous people (Lovera et al., 2017) across South America. This tuber is known for its pleasant taste (Pacheco et al., 2020), high starch content (up to 80%) (Chaali et al., 2020), and its high nutritional value. It contains highly digestible starch, as well as high levels of calcium, vitamin A (de Souza et al., 2021), ascorbic acid, protein, and fiber (Rincón Rueda et al., 2021), niacin, phosphorus, and iron (Londoño-Restrepo et al., 2018).

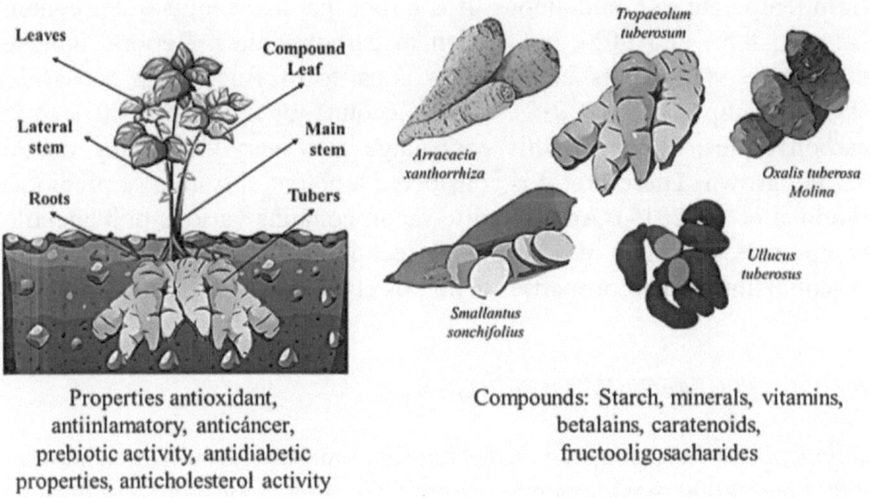

FIGURE 6.1 Tubers, compounds, and properties.

6.2.1.2 *OXALIS TUBEROSA MOLINA*

Oxalis tuberosa, known by various names depending on the region where it is cultivated, such as oca (Peru, Ecuador, and Bolivia), cuiba or quiba (Venezuela), ibia (Colombia), truffette acide (France), or foreign potato in Mexico (Morillo et al., 2019). It is cultivated at altitudes between 2,800 and 4,000 masl and is a key ingredient in the diets of the communities where it is grown. It is primarily consumed cooked (Clavijo-Ponce, 2018) and has a sweet taste with a floury texture (Morillo et al., 2019). This crop has a high-water content when fresh (around 80%) and is considered a source of carbohydrates, as it contains a percentage of sugars ranging from 3.6% to 21%, along with high levels of calcium and iron (Campos et al., 2018), which vary depending on the variety. Additionally, it has been reported to contain high concentrations of oxalic acid, proteins, and minerals (Morillo et al., 2019).

6.2.1.3 SMALLANTUS SONCHIFOLIUS

Smallantus sonchifolius, commonly known as yacón, it is cultivated between 1,800 and 2,800 masl, in regions ranging from southern Colombia to northern Argentina, it adapts to different climates and soils, from temperate to mountainous. It is a root that has a high-water content between 83% and 90% and is rich in carbohydrates. Reports indicate that up to 90% of its composition consists of fructooligosaccharides (FOS) (Campos et al., 2018), which account for more than 50% of its carbohydrates. However, this percentage may vary depending on the variety grown. Therefore, it is considered a potential source of prebiotics (Padilha et al., 2017). Additionally, yacón contains various polyphenolic compounds that make it a promising technological alternative due to its potential antioxidant properties (Campos et al., 2018).

6.2.1.4 TROPAEOLUM TUBEROSUM

Tropaeolum tuberosum, commonly known as mashua, is cultivated between 2,800 and 4,000 masl, extending from Colombia to Bolivia, and presents variations on its shape and color (white, yellow, purple or red) (Campos et al., 2018). It is the fourth most important Andean tuber, recognized for its nutritional and medicinal value (Heinert et al., 2021). *T. tuberosum* tolerates cold temperatures and is resistant to different pests, thanks to its high content of glucosinolates. This tuber is rich in carbohydrates, protein, sugars, and fiber, while being low in fat. Additionally, it contains various carotenoids and bioactive compounds such as anthocyanins (Campos et al., 2018). Due to its chemical composition, this crop has been attributed with antimicrobial, anti-inflammatory, and antioxidant properties (Silva-Correa et al., 2022).

6.2.1.5 ULLUCUS TUBEROSUS

Ullucus tuberosus is the second most cultivated tuber in the South American Andes and is only surpassed by potato cultivation. It is typically consumed as a vegetable to accompany meats, and its common name varies depending on the region of production: melloco (Ecuador), ulluco (Peru), chugua (Colombia), or ruba (Venezuela), among others (Cejudo-Bastante et al., 2014). The shape of this vegetable can be cylindrical, ovoid, or spherical, and its colors range from white, yellow, yellow with red spots, pink, orange, magenta, to red, depending on the variety. The coloration is attributed to a natural and functional pigment called betalain (Carlos et al., 2021; Pacheco et al., 2019, 2020). It contains approximately 65% starch on a dry basis and is a potential source of protein, fiber, vitamin C, and betalains (Repo-Carrasco & Cortez, 1988; Heil et al., 2017; Campos et al., 2018; Pacheco et al., 2020; Mosquera et al., 2020). The nutritional value of this vegetable varies depending on the variety, but it represents a potential source of bioactive compounds. Certain health effects have been attributed to this crop, including the reduction of inflammatory processes and regenerative properties with procollagen (Galindez et al., 2019).

6.2.2 NUTRITIONAL COMPOSITION OF ANDEAN TUBERS

The nutritional composition of some Andean tubers is shown in Table 6.1, where the starch content varies according to the species and variety produced (King & Gershoff, 1987). The protein content can vary between 0.7% and 8.5%, lipid content values between 0.1 and 1.1%, a fiber content of 1% to 9%, and the main carbohydrate present is starch with values between 60% and 90%, approximately.

6.2.3 ASSESSMENT OF ANDEAN TUBERS

There is growing concern in the Andean region regarding the losses incurred in the handling of these tubers after harvest, due to their underutilization (Campos et al., 2018). However, it is known that these tubers possess bioactive compounds, nutritional properties, and functional value due to the presence of secondary metabolites (Table 6.2) (Silva-Correa et al., 2022). Additionally, the inhabitants of the region where they are produced mainly consume them fresh (Velásquez-Barreto et al., 2021). Therefore, in order to fully utilize the biological potential of these tubers, it is necessary to expand the alternatives for their production, application, and commercialization (Galíndez et al., 2019). Furthermore, it is important to showcase and disseminate an overview of the bioactive compounds (antioxidants) they contain, as well as their nutritional composition (fats, carbohydrates, proteins, fiber, and minerals).

TABLE 6.1 Proximal Composition of Various Andean Tubers

Tuber	Humid (%)	Protein (%)	Oil (%)	Fiber (%)	Total Fiber (%)	Ash (%)	Minerals	Vitamins	References
Arracacia xanthorrhiza	–	0.7%	0.5%	86%	1.1%	15%	Calcium	Vitamin A Niacin	Leidi et al. (2018); Güçlü-Üstündağ y Mazza (2007); Tribst et al. (2016); de Souza et al. (2021)
Oxalis tuberosa Molina	14%	0.52%	0.15%	90.5%	–	0.24%	Calcium Iron	–	Moscoe et al. (2017); Hernández-Lauzardo et al. (2004); Campos et al. (2018)
Smallantus sonchifolius	6%	2.7%	0.17%	82.72%	9%	2.8%	–	–	Fabersani et al. (2018); Leidi et al. (2018)
Tropaeolum tuberosum	–	4.2%	1.1%	78%	6.8%	0.6%	–	Vitamin C	Pacheco et al. (2020); Leidi et al. (2018); Betalleluz-Pallardel et al. (2012).
Ullucus tuberosus	7.8%	8.5%	0.1%	64.96%	1.4%	–	–	Vitamin C	Pacheco et al. (2019); Velásquez-Barreto et al. (2019); Heil et al. (2017); Campos et al. (2018); Mosquera et al. (2020)

TABLE 6.2 Secondary Metabolites in Andean Tubers

Tuber	Compound	General Structure	Description	References
Smallantus sonchifolius	Fructooligosaccharides	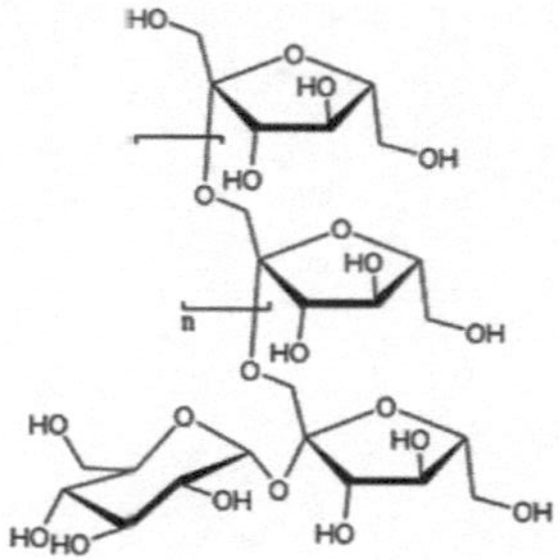	The oligosaccharides obtained by the hydrolysis of inulin, through endo-inulinases or enzymatic transfructosylation (β-frutosyltransferase), sucrose being the substrate.	Paredes et al. (2018); Nicolás et al. (2021)
Oxalis tuberosa Molina	Anthocyanins • Malvidin (malvidin 3-O-glucoside and 3,5-O-diglucoside) • Petunidin • Peonidin • Delphinidin	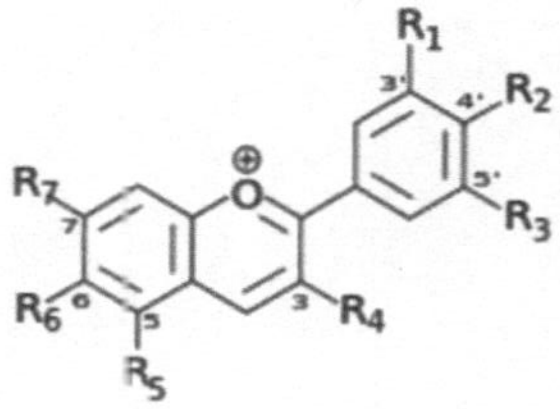	Some glycosides that come from anthocyanidins, they have a red coloration, and, in their structure, they have an anthocyanidin molecule, which is the aglycone, to which a sugar is attached through a β-glycosidic bond.	Alcalde-Eon et al. (2004); Miguel et al. (2011)
	Phenolic compounds • Caffeic acid • Cinnamic acid • Flavones and derivatives (flavan-3-ols)	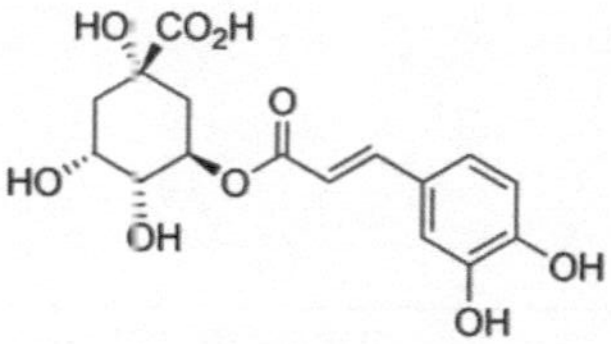	The chemical structures that have in their structure an aromatic ring attached to a hydroxyl group and are distributed in the plant kingdom.	Chirinos et al. (2009)

TABLE 6.2 *(Continued)*

Tuber	Compound	General Structure	Description	References
Ullucus tuberosus	Betalains • Betaxanthins (histidine-betaxanthin, arginine-betaxanthin, and glutamine-betaxanthin) • Betacyanins (betanin/ isobetanin)	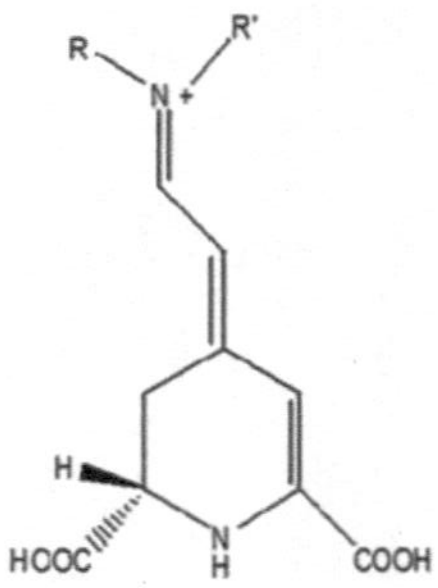	They are water-soluble pigments of bright colors, they have betalamic acid in the main structure, which in its condensation with an amine or amino acid produces betaxanthins, and condensed with cyclo-dihydroxyphenylalanine generates betacyanines.	Quillay-Davila et al. (2018)
Arracacia xanthorrhiza	Glycosics • Saponins	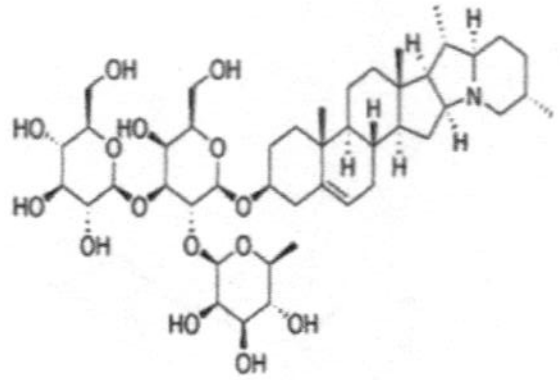	The glycosidic compounds, which present in their general formula (CnH2n-810). They can have a skeleton of triterpenic type (C_{30}) or steroidal type (C_{27}).	Quillay-Davila et al. (2018)
	Fructooligosaccharides	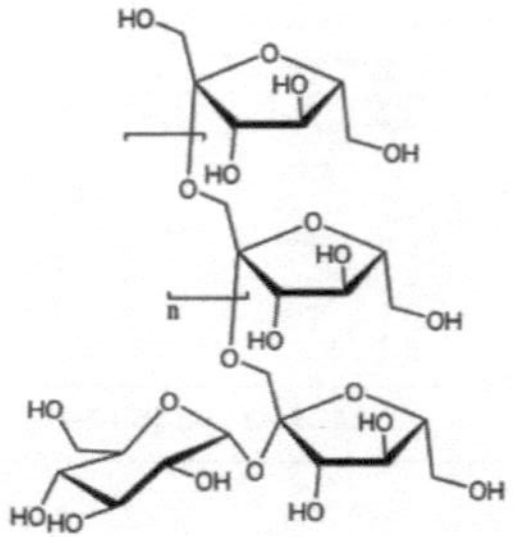	These are oligosaccharides obtained by the hydrolysis of inulin, through endo-inulinases or enzymatic transfructosylation (β-frutosyltransferase), sucrose being the substrate.	Lovera et al. (2017); Nicolás et al. (2021)

TABLE 6.2 *(Continued)*

Tuber	Compound	General Structure	Description	References
	Carotenoids	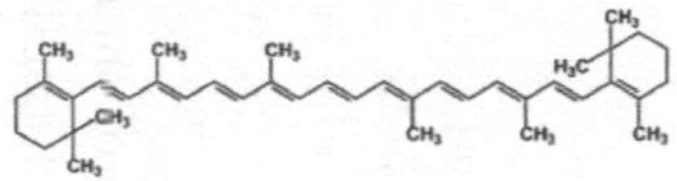	They are naturally synthesized fat-soluble terpenes and are made up of several isoprene units with a substituted and unsaturated cyclohexane ring at each end. These are found in plants, algae, and photosynthetic bacteria.	Güçlü-Üstündağ y Mazza (2007); Tribst et al. (2016); Elena et al. (2011)
Tropaeolum tuberosum	Phenolic compounds • Gallic acid • Gallocatechin • Procyanidin B2 • Epigallocatechin • Hydroxycinnamic and hydroxybenzoic acid derivatives • Derivatives of myricetin and proanthocyanidins	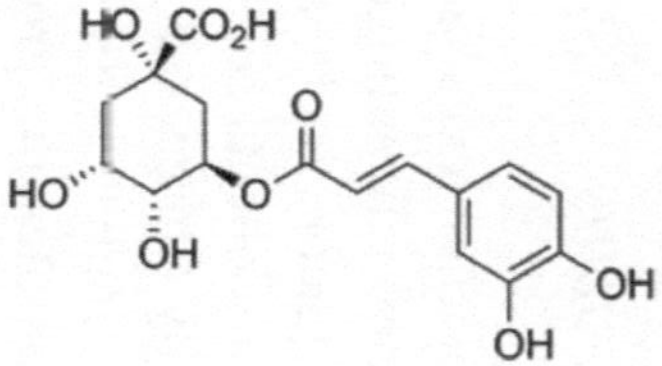	They are chemical structures that have in their structure an aromatic ring attached to a hydroxyl group and are distributed in the plant kingdom. These present biological activity within the organism, such as anticancer, anti-inflammatory and antioxidant activity, among others.	Betalleluz-Pallardel et al. (2012); Ortega et al. (2006); Martín Gordo (2018)
–	Glycosides • Glycosilanates: 4-hydroxybenzyl (glucosialbin), benzyl glucosinolate (glucotropaeoline), m-methoxybenzyl glucosinolate (glucolimnathin)	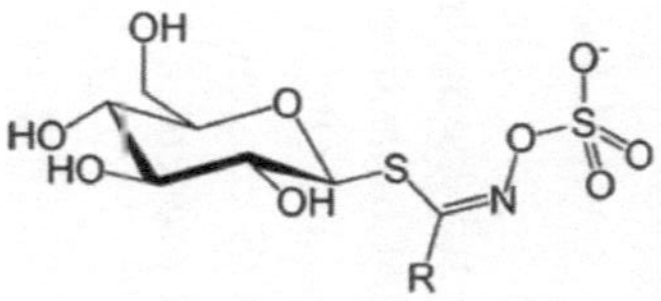	Naturally occurring in plants, they possess a glucosyl moiety in their structure (glycolyl "gluco,") a sulfate group, and are hydrolyzed by the enzyme myrosinase to biologically active compounds such as isothiocyanate sulfurans, nitriles, and thiocyanates.	Ortega et al. (2006); Rampal et al. (2009); Rincón Pérez (2014)

6.2.4 BIOACTIVE COMPOUNDS IN ANDEAN TUBERS

Some foods have biological activity that can produce beneficial health effects (López Carreras et al., 2012). These compounds are called bioactive and are found naturally, originating during the secondary metabolism of plants. They are also known as phytochemical compounds (Torres-León et al., 2017). These compounds (Table 6.3) are associated with the ability to produce beneficial effects on health and can help in the treatment of various diseases, such as those associated with some types of cancer (Ascacio-Valdés et al., 2016) and inflammation (Torres-León et al., 2017), given that they contain antioxidant compounds (García-Mayordonmo, 2016). Antioxidants are compounds that have the capacity to eliminate free radicals within the organism and, in turn, provide protection against oxidative reactions (Torres-León et al., 2017).

6.2.4.1 SMALLANTUS SONCHIFOLIUS

Andean tubers are recognized for their excellent nutritional value, their content of bioactive components, mainly antioxidants and FOS, and their prebiotic activity. In the case of yacón, its bioactive potential is related to its FOS content, which are components selectively used by probiotic bacteria present in the human body (Paredes et al., 2018). The fermentation of FOS by the gut microbiota increases the production of short-chain fatty acids, reduces pH, and promotes the growth of lactobacillus and bifidobacteria. Genta et al. (2009) conducted a study on the effect of yacón syrup consumption in women with obesity and dyslipidemia during the premenopausal period, and found that weight, waist circumference, and body mass index decreased. The consumption of FOS is associated with a delay in emptying time, leading to a prolonged feeling of satiety and reduced food intake (Rocha et al., 2018). Lower fasting serum insulin levels are also reported. In another study, Genta et al. (2005) evaluated the consumption of yacón flour in rats and observed a significant reduction in triglyceride levels, as well as a decrease in the development of chemically induced colon cancer. It is worth mentioning that most of these effects are attributed to the prebiotic effect; however, they may also be attributed to the beneficial action of phenolic compounds and fiber concentration (Gomes da Silva et al., 2017). The concentration of FOS, soluble sugars, and polysaccharides depends on the time of harvest and post-harvest treatment

(Lachman et al., 2004). Another effect of FOS consumption is the reduction of serum cholesterol, which can occur through different mechanisms such as lower blood glucose levels or the growth of cholesterol-assimilating strains of lactobacilli. Yacón leaves also contain bioactive compounds with antioxidant and anti-diabetic properties. Yacón leaf extracts are used for the prevention and treatment of chronic diseases, particularly diabetes and obesity. The concentration of bioactive compounds varies according to the cultivar (red or white), leaf age, and the extraction process used (Khajehei et al., 2017).

6.2.4.2 OXALIS TUBEROSA MOLINA

Different studies have evaluated the antioxidant capacity of phenols, carotenoids, anthocyanins, betaxanthin, and betacyanin of oca and ulluco by the ABTS test. In addition, the main bioactive compounds in *O. tuberosa* were phenols and anthocyanins. The main anthocyanins present in purple *O. tuberosa* were malvidin glycosides (malvidin 3-O-glucoside and 3,5-O-diglucoside), petunidin, peonidin, delphinidin (mono- and/or diglycosylates), and only one acylated anthocyanin (malvidin 3-O-acetyl-glucoside-5-O-glucoside) (Alcalde-Eon et al., 2004). The non-anthocyanin phenolic compounds present in six different *O. tuberosa* (yellow and purple) were caffeic and cinnamic acid derivatives, as well as flavan-3-ols, flavones, and flavanone derivatives. The flavan-3-ols, flavones, and flavanones detected corresponded to bound forms of catechin, luteolin, and apigenin and naringenin, respectively, according to their UV-Vis spectral data (Chirinos et al., 2009). It was also found that compounds from the aqueous fraction were the major contributors to the antioxidant capacity of the ABTS test.

6.2.4.3 ULLUCUS TUBEROSUS

Ulluco contains betalains in the form of betaxanthins and betacyanins. Betalains have been identified in red and yellow varieties of ulluco, notably histidine-betaxanthin, arginine-betaxanthin, and glutamine-betaxanthin. Betacyanins were found in red tubers, at approximately 50% betanin/isobetanin. Betacyanin levels were up to 70 $\mu g\ g^{-1}$ FW in red tubers, and betaxanthin levels were up to 50 $\mu g\ g^{-1}$ FW in yellow tubers (Campos et al., 2018; Márquez-Fernández et al., 2019). Likewise, new betalains (betacyanin and betaxanthin) were separated by fractions of *U. tuberosus* Caldas through HPLC-DAD-ESI-MS, where the chemical identification of each fraction of betalains was developed, revealing a large number of those that had not been described in previous investigations of these tubers. Some of these newly identified betalains include betanidin- and isobetanidin-5-O-(4′-O-malonyl-β-glucoside), 2-decarboxyphyllocactin, betanidin-, and isobetanidin-6-O-(6′-O-feruloyl)-β-glucoside (gomphrenin and isogomphrenin III), dehydrophyllocactin, and isophyllocactin, and arginine and glycine-betaxanthins (portulacaxanthin III). The quantification of these betalains and the antioxidant activity of betacyanin (44.5 μg betanin/g fresh ulluco) and betaxanthins (21.8 μg indicaxanthin/g fresh ulluco) were also determined (Mosquera et al., 2020). These results suggest that this tuber is a bioactive food source.

6.2.4.4 ARRACACIA XANTHORRHIZA

Arracacha, as well as yacón, has been mentioned to have a significant content of resistant starch and FOS (Lovera et al., 2017). This is linked to a lower glycemic index when consumed, as well as being considered as a substrate for the intestinal microbiota (Sancho et al., 2017). Regarding carotenoid content, an important source is provitamin A present in this root. The bioactive components in arracacha correspond to phenolic compounds with diuretic and antiseptic activity, in addition to saponins with anti-cancer and anti-cholesterol activity (Güçlü-Üstündağ & Mazza, 2007; Tribst et al., 2016).

Furthermore, research has been conducted on the structure and properties of starches from Arracacha roots (*Arracacia xanthorrhiza*), evaluating and characterizing starches obtained from two varieties of Peruvian arracacha (purple and yellow arracacha) in relation to granular morphology, molecular structure, and properties. Long amylopectin lengths were identified, which showed differences of up to 20% compared to other research. This makes this crop of great interest for industrial purposes. The characteristics presented were – low gelatinization temperature, gelatinization enthalpy, and tendency to retrogradation, high peak apparent viscosity and swelling capacity, even at moderate temperatures (60°C) (Castanha et al., 2018). These properties have allowed the starch obtained to be used in the food industry (baking) and offer the possibility of its application in functional foods (Scortegagna et al., 2020), since it is known that starch is the carbohydrate that provides the most energy (Luo et al., 2021), which can be used more easily for oxidative metabolism (Ojeda et al., 2018).

6.2.4.5 TROPAEOLUM TUBEROSUM

Mashua is an important source of vitamin C, antioxidants, and glucosinolates. The phenolic compounds found in this tuber include gallic acid, gallocatechin, procyanidin B2, epigallocatechin, flavan 3-ol derivatives, hydroxycinnamic and hydroxybenzoic acid derivatives, rutin, and/or myricetin derivatives, and proanthocyanidins. Proanthocyanidin fractions (prodelphinidins) were the main contributors to antioxidant activity. The main anthocyanins identified in mashua root were delphinidins, cyanidins, and pelargonidins. Additionally, mashua polyphenol extracts exhibited antioxidant properties (Betalleluz-Pallardel et al., 2012).

Regarding glucosinolate content, dark-colored tubercle varieties showed higher levels compared to light-colored varieties. The detected glucosinolates were 4-hydroxybenzyl glucosinolate (glucosialbin), benzyl glucosinolate (glucotropaeolin), and m-methoxybenzyl glucosinolate (glucolimnathin) (Ortega et al., 2006). Glucosinolates are a group of compounds produced during the secondary metabolism of plants, which can be classified as aliphatic, aromatic, or indole. These compounds are hydrolyzed by the myrosinase enzyme, breaking them down into different products such as isothiocyanates with antifungal, antibacterial, antioxidant, and anticancer properties (Rampal et al., 2009).

Several thiocyanate-rich extracts of mashua have shown anticancer activity against colon and prostate cancer cell lines (Martín & Higuera, 2016).

6.3 TECHNOLOGICAL PERSPECTIVE

Adding value to Andean tubers such as arracacha, oca, yacón, mashua, and ulluco can have a significant impact on the Andean communities where these crops are produced (Table 6.4). This highlights the need to characterize the chemical, biological, and structural properties of these tubers in order to propose technological applications and identify future trends for their use as raw materials.

TABLE 6.3 Bioactive Compounds and Properties of Andean Tubers

Tuber	Bioactive Component	Properties	References
Arracacia xanthorrhiza	Phenolic compounds, saponins.	Diuretic, antiseptic, anti-carcinogenic, and anti-cholesterol activity.	Güçlü-Üstündağ & Mazza (2007); Tribst et al. (2016)
Oxalis tuberosa Molina	Phenols and anthocyanins.	Antioxidant capacity	Chirinos et al. (2009); Alcalde-Eon et al. (2004)
Smallantus sonchifolius	Fructo-oligosaccharides (FOS)	Prebiotic effect	Paredes et al. (2018)
Tropaeolum tuberosum	Glycosylanates	Antioxidant properties and anticarcinogenic activity.	Martín & Higuera (2016); Betalleluz-Pallardel et al. (2012)
Ullucus tuberosus	Betalaines	Reduction of inflammatory processes and regenerative properties with procollagen. Antioxidant properties.	Galindez et al. (2019)

One of the most valuable components found in these tubers is starch. It is worth noting that Andean tubers are recognized for their starch content, which possesses desirable technological properties such as gelatinization at low temperatures, low retrogradation tendency, and high viscosity. This starch can be utilized in the food industry as a thickening and gelling agent to enhance the texture of various food products. Moreover, the starch present in these tubers is considered an excellent ingredient due to its technological properties, making it suitable for applications in the bakery, dairy, and cereal industries (Castanha et al., 2018).

Numerous studies have reported the excellent potential of Andean tubers and roots as fermented products, suggesting that they can be considered as functional foods due to their probiotic and prebiotic properties (Luziatelli et al., 2020). Additionally, these tubers serve as a source of alternative sugars. Furthermore, Andean tubers can be utilized in the formulation of new and improved foods for the vegetarian market, given their high protein, energy, and mineral content. The nutritional, bioactive, and functional properties of Andean tubers play a crucial role in the development of functional foods (Leidi et al., 2018).

TABLE 6.4 Principal Uses of Andean Tuber

Component	Application	Use	References
Starch	Food industry	Thickener and gelling agent.	Castanha et al. (2018)
Fructo-oligosaccharides		Functional foods potential as fermented products.	Luziatelli et al. (2020)
Protein, energy, and minerals		Vegetarian market.	Leidi et al. (2018)

6.4 CONCLUSION

Andean tubers represent an excellent source of nutritional and bioactive compounds. These characteristics mean that tubers are considered as an ingredient for the development of functional foods. In addition to the above, their technological characteristics are applicable to the current needs of the food industry. The Andean region is characterized by its biodiversity; therefore, Andean tubers show a marked difference between species, especially in terms of nutrients and bioactive compounds. Over the years, efforts have been made to characterize and determine their main bioactive properties. However, it is necessary to invest in research on these tubers to determine their *in vivo* effect. The use of these foods outside the Andean region is very limited, so strategies must be developed to awaken the interest of other markets to encourage their consumption and, at the same time, increase studies on their properties as functional and technological nutrients applied in the food industry.

Despite the fact that some research has evaluated the properties of these tubers, which have a wide potential for application (Barrera et al., 2003), research has not been very extensive and little information is currently available, so there is still limited scientific knowledge (Ponce & Martínez, 2014) about these tubers. This topic is an area of opportunity for further research, which would allow the nutritional, bioactive, and functional potential of the Andean tubers in question to be exploited and promoted.

KEYWORDS

- **Andean tubers**
- **arracacha**
- **bioactive compounds**
- **food industry**
- **gelatinization**
- **mashua**
- **oca**
- **retrogradation tendency**
- **ulluco**
- **yacón**

REFERENCES

Andrade-Mahecha, M. M., Tapia-Blácido, D. R., & Menegalli, F. C., (2012). Physical-chemical, thermal, and functional properties of achira (*Canna indica L.*) flour and starch from different geographical origin. *Starch/Staerke*, *64*(5), 348–358. https://doi.org/10.1002/star.201100149.

Ascacio-Valdés, J. A., Aguilera-Carbó, A. F., Buenrostro, J. J., Prado-Barragán, A., Rodríguez-Herrera, R., & Aguilar, C. N., (2016). The complete biodegradation pathway of ellagitannins by *Aspergillus Niger* in solid-state fermentation. *Journal of Basic Microbiology*, *56*(4), 329–336. https://doi.org/10.1002/jobm.201500557.

Betalleluz-Pallardel, I., Chirinos, R., Rogez, H., Pedreschi, R., & Campos, D., (2012). Phenolic compounds from Andean mashua (*Tropaeolum tuberosum*) tubers display protection against soybean oil oxidation. *Food Science and Technology International*, *18*(3), 271–280.

Campos, D., Chirinos, R., Gálvez, R. L., & Pedreschi, R., (2018). Bioactive potential of Andean fruits, seeds, and tubers. *Advances in Food and Nutrition Research*, 287–343. doi: 10.1016/bs.afnr.2017.12.005.

Castanha, N., Villar, J., Matta, Jr. M. D. Da., Anjos, C. B. P. D., & Augusto, P. E. D., (2018). Structure and properties of starches from arracacha (*Arracacia xanthorrhiza*) roots. *International Journal of Biological Macromolecules, 117*, 1029–1038. doi: 10.1016/j.ijbiomac.2018.06.015.

Cejudo-Bastante, M. J., Hurtado, N., Mosquera, N., & Heredia, F. J., (2014). Potential use of new Colombian sources of betalains. Color stability of ulluco (*Ullucus tuberosus*) extracts under different pH and thermal conditions. *Food Research International*, *64*, 465–471. https://doi.org/10.1016/j.foodres.2014.07.036.

Chaali, N., Ouazaa, S., Jaramillo-Barrios, C. I., Araujo, C. G. A., & Ávila, P. E. Á., (2020). Edaphoclimatic characterization and crop water requirement of arracacha (*Arracacia xanthorrhiza Bancroft*) roots in upland production areas. *Scientia Horticulturae, 272*. https://doi.org/10.1016/j.scienta.2020.109533.

Clavijo, P. N. L., (2018). Cultura y conservación in situ de tubérculos andinos marginados en agroecosistemas de Boyacá: Un análisis de su persistencia desde la época prehispánica hasta el año 2016. *Cuadernos de Desarrollo Rural*, *14*(80). https://doi.org/10.11144/javeriana.cdr14-80.ccst.

Cruz-Tirado, J. P., Vejarano, R., Tapia-Blácido, D. R., Barraza-Jáuregui, G., & Siche, R., (2019). Biodegradable foam tray based on starches isolated from different *Peruvian* species. *International Journal of Biological Macromolecules*, *125*, 800–807. https://doi.org/10.1016/j.ijbiomac.2018.12.111.

Duque, L., Bravo, K., & Osorio, E., (2017). A holistic anti-aging approach applied in selected cultivated medicinal plants: A view of photoprotection of the skin by different mechanisms. *Industrial Crops and Products*, *97*, 431–439. https://doi.org/10.1016/j.indcrop.2016.12.059.

Fabersani, E., Grande, M. V., Coll, A. M. V., Zannier, M. L., Sánchez, S. S., Grau, A., Oliszewski, R., & Honoré, S. M., (2018). Metabolic effects of goat milk yogurt supplemented with yacon flour in rats on high-fat diet. *Journal of Functional Foods*, *49*, 447–457. https://doi.org/10.1016/j.jff.2018.08.042.

Galindez, A., Daza, L. D., Homez-Jara, A., Eim, V. S., & Váquiro, H. A., (2019). Characterization of ulluco starch and its potential for use in edible films prepared at low drying temperature. *Carbohydrate Polymers*, *215*, 143–150. https://doi.org/10.1016/j.carbpol.2019.03.074.

Genta, S. B., Cabrera, W. M., Grau, A., & Sanchez, S. S., (2005). Subchronic 4-month oral toxicity study of dried *Smallanthus sonchifolius* (yacon) roots as a diet supplement in rats. *Food and Chemical Toxicology*, *43*, 1657–1665.

Genta, S., Cabrera, W., Habib, N., Pons, J., Manrique-Carillo, I., Grau, A., et al., (2009). Yacon syrup: Beneficial effects on obesity and insulin resistance in humans. *Clinical Nutrition*, *28*, 182–187.

Goicochea, R. C. C., Carolim, M. M. L., & Teodoro, J. R. C., (2021). Propiedades funcionales de productos tradicionales congelados y secados al sol de oca (*Oxalis tuberosa Molina*) y olluco (*Ullucus tuberosus Caldas*): Una revisión. *Puriq*, *2*(3).

Gomes Da, S. M. F., Dionísio, A. P., Ferreira, C. A. A., Silveira, A. L., Oliveira, P. C., Pinto De, A., F. A., & Ferreira, P. D., (2017). Yacon syrup: Food applications and impact on satiety in healthy volunteers. *Food Research International*, *100*, 460–467.

Güçlü-Üstündağ, Ö., & Mazza, G., (2007). Saponins: Properties, applications, and processing. *Critical Reviews in Food Science and Nutrition*, *47*, 231–258.

Heil, N., Bravo, K., Montoya, A., Robledo, S., & Osorio, E., (2017). Wound healing activity of *Ullucus tuberosus*, an Andean tuber crop. *Asian Pacific Journal of Tropical Biomedicine*, *7*(6), 538–543. https://doi.org/10.1016/j.apjtb.2017.05.007.

Khajehei, F., Niakousari, M., Damyeh, M. S., Merkt, N., Claupein, W., & GraeffHoenninger, S., (2017). Impact of ohmic-assisted decoction on bioactive components extracted from yacon (*Smallanthus sonchifolius Poepp.*) Leaves: Comparison with conventional decoction. *Molecules, 22, 2043.* https://doi.org/10.3390/ molecules22122043.

King, S. R., & Gershoff, S. N., (1987). Nutritional evaluation of three underexploited Andean tubers: *Oxalis tuberosa* (*Oxalidaceae*), *Ullucus tuberosus* (*Basellaceae*), and

Tropaeolum tuberosum (*Tropaeolaceae*). *Economic Botany*, *41*(4), 503–511. https://doi.org/10.1007/BF02908144.

Lachman, J., Havrland, B., Fernández, E. C., & Dudjak, J., (2004). Saccharides of yacon [*Smallanthus sonchifolius* (Poepp. et Endl.) H. Robinson] tubers and rhizomes and factors affecting their content. *Plant, Soil and Environment*, *50*, 383–390.

Londoño-Restrepo, S. M., Rincón-Londoño, N., Contreras-Padilla, M., Millan-Malo, B. M., & Rodriguez-Garcia, M. E., (2018). Morphological, structural, thermal, compositional, vibrational, and pasting characterization of white, yellow, and purple arracacha Lego-like starches and flours (*Arracacia xanthorrhiza*). *International Journal of Biological Macromolecules*, *113*, 1188–1197. https://doi.org/10.1016/j.ijbiomac.2018.03.021.

López, C. N., Miguel, M., & Aleixandre, A., (2012). Propiedades beneficiosas de los terpenos iridoides sobre la salud. *Nutricion Clinica y Dietetica Hospitalaria*, *32*(3), 81–91.

Lovera, M., Pérez, E., & Laurentin, A., (2017). Digestibility of starches isolated from stem and root tubers of arracacha, cassava, cush–cush yam, potato, and taro. *Carbohydrate Polymers*, *176*, 50–55. https://doi.org/10.1016/j.carbpol.2017.08.049.

Luziatelli, G., Sørensen, M., & Jacobsen, S. E., (2020). Current uses of Andean roots and tuber crops in south american gourmet restaurants. *International Journal of Gastronomy and Food Science, 100270.* doi: 10.1016/j.ijgfs.2020.100270.

María, B., Carlos, U., R. G., & P. Y., (2016). Estudio de cuatro tubérculos y raíces tuberosas no tradicionales De La Sierra Centro De Ecuador Y Su. *Escuela de Administración de Empresas Gastronómicas*, *12*, 37–67. https://www.unibe.edu.ec/wp-content/uploads/2017/08/03_12_BONETEetal_RTAs-1.pdf (accessed on 24 February 2024).

Márquez, F. P. M., Fer, M., & Nández, D. M., (2019). Functional foods: Current issues and trends. *Vitae*, *26*(1), 6, 7. https://doi.org/10.17533/udea.vitae.v26n1a01.

Martı´n, J. C., & Higuera, B. L., (2016). Glucosinolate composition of Colombian accessions of mashua (*Tropaeolum tuberosum Ruíz & Pavón*), structural elucidation of the predominant glucosinolate and assessment of its antifungal activity. *Journal of the Science of Food and Agriculture*, *96*, 4702–4712.

Morillo C., A. C., Morillo C., Y., & Leguizamo M., & M. F., (2019). Caracterización morfológica y molecular de *Oxalis tuberosa Mol.* en el departamento de Boyacá. *Revista Colombiana de Biotecnología*, *21*(1), 18–28. https://doi.org/10.15446/rev.colomb.biote.v21n1.57356.

Mosquera, N., Cejudo-Bastante, M. J., Heredia, F. J., & Hurtado, N., (2020). Identification of new betalains in separated betacyanin and betaxanthin fractions from Ulluco (*Ullucus tuberosus Caldas*) by HPLC-DAD-ESI-MS. *Plant Foods for Human Nutrition*, *75*(3), 434–440. https://doi.org/10.1007/s11130-020-00837-9.

Ortega, O., Kliebenstein, D., Arbizu, C., Ortega, R., & Quiros, C., (2006). Glucosinolate survey of cultivated and feral mashua (*Tropaeolum tuberosum Ruiz & Pavon*) in the Cuzco region of Peru. *Economic Botany*, *60*, 254–264.

Pacheco, M. T., Hernández-Hernández, O., Moreno, F. J., & Villamiel, M., (2020). Andean tubers grown in Ecuador: New sources of functional ingredients. *Food Bioscience*, *35*. https://doi.org/10.1016/j.fbio.2020.100601.

Pacheco, M. T., Moreno, F. J., Moreno, R., Villamiel, M., & Hernandez-Hernandez, O., (2019). Morphological, technological, and nutritional properties of flours and starches from mashua (*Tropaeolum tuberosum*) and melloco (*Ullucus tuberosus*) cultivated in Ecuador. *Food Chemistry*, *301*, 125268. https://doi.org/10.1016/j.foodchem.2019.125268.

Padilha, V. M., Andrade, S. A. C., Valencia, M. S., Stamford, T. L. M., & Salgado, S. M., (2017). Optimization of synbiotic yogurts with yacon pulp (*Smallanthus sonchifolius*) and assessment of the viability of lactic acid bacteria. *Food Science and Technology, 37*(2), 166–175. https://doi.org/10.1590/1678-457X.14016.

Paredes, L. L. R., Smiderle, F. R., Santana-Filho, A. P., Kimura, A., Iacomini, M., & Sassaki, G. L., (2018). Yacon fructans (*Smallanthus sonchifolius*) extraction, characterization, and activation of macrophages to phagocyte yeast cells. *International Journal of Biological Macromolecules, 108*, 1074–1081.

Parra-Quijano, M., Panda, S., Rodríguez, N., & Torres, E., (2012). Diversity of *Ullucus tuberosus* (*Basellaceae*) in the Colombian Andes and notes on ulluco domestication based on morphological and molecular data. *Genetic Resources and Crop Evolution, 59*(1), 49–66. https://doi.org/10.1007/s10722-011-9667-8.

Paula, G. M., (2016). Compuestos bioactivos en alimentos de origen vegetal. *Universidad Complutense*, 1–20.

Quijano-Célis, C. E., Piedrahita, D., & Pino, J. A., (2016). Essential oil of *Arracacia xanthorrhiza Bancr.* leaves from Colombia. *Journal of Essential Oil-Bearing Plants, 19*(5), 1296–1299. https://doi.org/10.1080/0972060X.2016.1195710.

Quillay-Davila, M., Arana-Arias, Y., Jaramillo-Jaramillo, C., Cuenca-Buele, S., Rojas-De-Astudillo, L., & Jaramillo-Alcívar, V., (2018). Contenido de saponinas y actividad cicatrizante de Cecropia peltata y Parthenium hysterophorus. *Revista Cubana de Farmacia [Internet], 51*(3). Disponible en: http://www.revfarmacia.sld.cu/index.php/far/article/view/250 (accessed on 24 February 2024).

Repo-Carrasco, R., Cortez, G., Montes, R. O., & Villalpando, L. Q., (1988). Cultivos andinos. *Importancia Nutricional y Posibilidades de Procesamiento.* Cusco: Bartolomé de La Casas.

Rincón, R. M. A., Ruiz, B. H. D., Molano, D. J. M., Alvarez, H. J. G., & Pinto, A. Y. L., (2021). Postharvest characterization of seven arracacha cultivars (*Arracacia xanthorrhiza Bancroft*). *Revista Facultad Nacional de Agronomía Medellín, 74*(3), 9745–9756. https://doi.org/10.15446/rfnam.v74n3.92658.

Rocha, D. M. U. P., Ribeiro, P. V. M., Caldas, A. P. S., Silva, B. P., Silva, A., Almeida, A. P., & Alfenas, R. C. G., (2018). Acute consumption of yacon shake did not affect glycemic response in euglycemic, normal weight, healthy adults. *Journal of Functional Foods, 44*, 58–64.

Salazar, D., Arancibia, M., Ocaña, I., Rodríguez-Maecker, R., Bedón, M., López-Caballero, M. E., & Montero, M. P., (2021). Characterization and technological potential of underutilized ancestral Andean crop flours from Ecuador. *Agronomy, 11*(9), 1693. https://doi.org/10.3390/agronomy11091693.

Sancho, R. A. S., Souza, J. D. R., De Lima, F. A., & Pastore, G. M., (2017). Evaluation of oligosaccharide profiles in selected cooked tubers and roots subjected to in vitro digestion. *LWT-Food Science and Technology, 76*, 270–277.

Torres-León, C., Rojas, R., Serna-Cock, L., Belmares-Cerda, R., & Aguilar, C. N., (2017). Extraction of antioxidants from mango seed kernel: Optimization assisted by microwave. *Food and Bioproducts Processing, 105*, 188–196. https://doi.org/10.1016/j.fbp.2017.07.005.

Tribst, A. A. L., De Castro, L. B. R. Jr., De O., Miguel, M., & Cristianini, M., (2016). High pressure processing of cocoyam, Peruvian carrot, and sweet potato: Effect on oxidative enzymes and impact in the tuber color. *Innovative Food Science and Emerging Technologies, 34*, 302–309.

Velásquez-Barreto, F. F., Bello-Pérez, L. A., Nuñez-Santiago, C., Yee-Madeira, H., & Velezmoro, S. C. E., (2021). Relationships among molecular, physicochemical and digestibility characteristics of Andean tuber starches. *International Journal of Biological Macromolecules, 182*, 472–481. https://doi.org/10.1016/j.ijbiomac.2021.04.039.

Vig, A. P., Rampal, G., Thind, T. S., & Arora, S., (2009). Bio-protective effects of glucosinolates. A review. *LWT – Food Science and Technology*, *42*, 1561–1572.

CHAPTER 7

Sustainable Valorization of Wheat Bran Waste Through the Extraction of Gelling Ferulated Arabinoxylan Intended for Encapsulation and Controlled Release of Metronidazole

OLYMPIA JUDITH DOMÍNGUEZ MURUA,[1,2]
ELIZABETH CARVAJAL-MILLAN,[1] JORGE A. MÁRQUEZ-ESCALANTE,[1]
AGUSTÍN RASCÓN CHU,[1] ALMA C. CAMPA-MADA,[1]
MARTIN ANTONIO ENCINAS ROMERO,[2] PAUL ZAVALA RIVERA,[2] and
AGUSTÍN GÓMEZ ÁLVAREZ[2]

[1]*Research Center for Food and Development (CIAD, A.C), Carretera Gustavo Enrique Astiazaran Rosas, Hermosillo, Sonora, Mexico*

[2]*Department of Chemical Engineering and Metallurgy, University of Sonora, Rosales y Blvd. Luis D. Colosio, Hermosillo, Sonora, Mexico*

ABSTRACT

Arabinoxylans (AX) are polysaccharides that constitute a structural part of the cell wall of cereals. These polysaccharides are formed by a linear chain of xyloses with arabinose branches to which ferulic acid may be esterified. The presence of ferulic acid gives AX the ability to create covalent gels through oxidation by enzymes, such as laccase. This oxidation results in the formation of dimers of ferulic acid (di-FA) that allow the polysaccharide chains to join, forming a gel. AX gels are neutral,

Sustainable Environment and Health: Practical Strategies. Leonardo Sepúlveda Torre, Ann Rose Abraham, Neha Kanwar Rawat, & A. K. Haghi (Eds.)

colorless, and odorless. Due to their covalent nature, they are not affected by changes in temperature, ionic strength, or pH, which gives them excellent application potential as encapsulation and controlled release agents. The objective of this research was to evaluate the ability of wheat bran AX gels to encapsulate and release metronidazole (MTZ) as a model molecule in a controlled manner. The AX formed 5% (w/v) covalent gels with an elasticity of 175 Pa, which decreased to 164 Pa in the presence of MTZ. The di-FA content in AX gels with and without MTZ was 0.01 μg/mg. The AX gel presents a microstructure with small and continuous porosities, while the AX/MTZ gel shows a less structured material with larger porosities. The apparent diffusion coefficient (Dm) of the MTZ released from the AX gel was 3.95×10^{-7} cm^2/s. In a time of 5 hours, only 18% of the encapsulated MTZ was released, which would allow the delivery of most of the drug loaded in the gel once the polymeric network is fermented by the colonic microbiota. The results indicate that AX gels can trap MTZ, forming matrices that could be used for the controlled release of this molecule.

7.1 INTRODUCTION

Arabinoxylans (AX) are neutral, non-starch polysaccharides found in cereal grains. They consist of a linear skeleton of β-(1-4)-D-xylanopyranose units with α-L-arabinofuranose substituents attached at O^{2-} and O^{3-} (Izydorczyk & Biliaderis, 1995). Some of the arabinose residues have esterified ferulic acid residues at $(O)^{5-}$ (3-methoxy, 4-hydroxycinnamic acid) (Smith & Hartley, 1983). Depending on their molecular weight (MW), AX can be soluble (MW<1,000 kDa) or insoluble (MW>1,000 kDa) in water (Izydorczyk & Biliaderis, 1995). The water-soluble AX is present in the endosperm of cereals, while the water-insoluble AX is mainly found in the bran of the grain. Due to their high molecular weight, the aqueous extraction of the water-insoluble AX involves partial hydrolysis of the molecule. The amount of water-insoluble AX in cereal bran is high (30–40% w/w) compared to the amount of water-soluble AX in the endosperm (1–3% w/w) (Carvajal-Millan et al., 2005a, 2007). AX can form gels under the oxidative action of enzymatic types (laccase, peroxidase/H_2O_2) on ferulic acid, resulting in the formation of free radicals. These free radicals then couple, leading to the formation of ferulic acid dimers (di-AF), which bind to polysaccharide chains and form a three-dimensional aqueous gel (Izydorczyk & Biliaderis, 1995; Figueroa-Espinoza & Rouau, 1998). To date, five isomeric forms of di-FA have been identified in AX gels (5-5′, 8-5′-benzo, 8-O-4′, 8-5′, and 8-8′). The 8-5′ and 8-O-4′ forms are generally more abundant in wheat AX gels (Figueroa-Espinoza & Rouau, 1998; Vansteenkiste et al., 2004; Carvajal-Millan et al., 2005a, b, 2006). However, in maize arabinoxylan gels, di-FA 5-5′ has been reported as the most abundant (Carvajal-Millan et al., 2007; Niño-Medina et al., 2009; Berlanga-Reyes et al., 2009). Until now, the formation of a single trimer of ferulic acid (tri-AF) (4-O-8′, 5′-5″-dehydrotriferulic acid) in arabinoxylan gels from wheat and corn has been reported (Carvajal-Millan et al., 2005a, 2007) induced by a laccase. In addition to the participation of the di-FA mentioned above and tri-AF structures, some authors point out the contribution of physical interactions between the AX chains' linear regions in establishing the gel's three-dimensional structure (Vansteenkiste et al., 2004; Carvajal-Millan et al., 2005b). AX gels have interesting characteristics: they are neutral, colorless, and odorless, and due to the presence of covalent bonds, they are little affected by changes in temperature, ionic strength, or pH (Izydorczyk & Biliaderis, 1995; Niño-Medina et al., 2010; Morales-Burgos et al., 2017). These characteristics, along with the meso- and macroporous structure of AX gels with a pore size ranging from 40 to 400 nm, offer great potential for AX to be used as a colon-targeted controlled drug release system (Carvajal-Millan et al., 2005b; Berlanga-Reyes et al., 2014; Martinez Lopez et al., 2016). Metronidazole (MTZ) is a synthetic antibacterial and antiparasitic agent classified as a nitroimidazole (Bendesky & Menéndez, 2012). MTZ can be administered intravenously, topically, vaginally, and rectally, although oral administration is the most common. This drug is widely used for the oral treatment of intestinal amebiasis. However, it exhibits high permeability in the upper gastrointestinal system, which hinders the delivery of adequate doses to the site of infection, the colon. Therefore, there is significant interest in developing alternatives for oral delivery of MTZ, with polymeric encapsulation showing great potential for delivering this drug to the large intestine, particularly the colon region. The present research aims to evaluate the ability of AX gels derived from wheat bran to encapsulate and release MTZ in a controlled manner, serving as a model for antiparasitic molecules.

7.2 MATERIALS AND METHODS

7.2.1 RAW MATERIAL

Wheat bran AX, previously extracted and characterized using methods developed by the CIAD Biopolymer Research Group, was utilized. The AX exhibited an A/X ratio of 0.80, a molecular weight of 74 kDa, an [ŋ] of 206 mL/g, and a ferulic acid content of 0.02 μg/mg AX (Berlanga-Reyes et al., 2011). A commercial laccase (*Trametes versicolor*) was employed as a gelling agent. Commercial MTZ was also utilized.

7.2.2 PREPARATION OF AX GELS WITH MTZ

Solutions of 5% (w/v) AX and 20% MTZ were prepared using 0.5 M citrate phosphate buffer with a pH of 5.5. Gels were formed by combining 2 mL of AX/MTZ solution with 50 μL of laccase (1,675 nkat/mg of AX). Gelation was conducted for 1.5 hours at 25°C.

7.2.3 CONTENT OF FERULIC ACID, ITS DIMERS, AND TRIMERS

It was performed using reverse-phase high-performance liquid chromatography (RP-HPLC). The phenolic compounds were extracted through controlled saponification and recovery in ether, following the method described by Vansteenksite et al. (2004). To 0.5 mL of AX gel, 1 mL of 1N NaOH was added in an argon-enriched atmosphere and incubated for 2 hours in the dark at 35°C. Subsequently, 10 μg of 3,4,5-trimethoxy-trans-cinnamic acid (TMCA) was added as an internal standard, and the pH was adjusted to 2.0 ± 0.2 with 4N HCl. The phenols were extracted twice with diethyl ether and evaporated at 30°C in the presence of argon. The dry extract was dissolved in 0.50 mL of a methanol/water/acetic acid mixture (40/59/01), filtered (0.45 μm), and injected (20 μL) into a Supelcosil column (LC-18-DB 250 × 4.6 mm, Supelco, Inc., Bellefonte, USA). The column temperature was maintained at 35°C, and isocratic elution was achieved using a methanol/water/acetic acid (40/59/01) mobile phase at a flow rate of 0.6 mL/min. The equipment used was a Waters 996 (Millipore Co., Milford, USA) equipped with a diode array detector. Detection was performed at 280 nm.

7.2.4 *DYNAMIC RHEOLOGY*

Rheological measurements were conducted using a small deformation dynamic rheometer (Discovery HR-2 rheometer; TA Instruments, New Castle, USA) following the method described by Vansteenkiste et al. (2004). The gelation kinetics of the AX were recorded for 6 hours at 25°C. Measurements were performed at a frequency of 0.25 Hz and a strain of 5%. A deformation scan of the gel was carried out, revealing that the region of linear behavior occurs between 1.5% and 10% deformation. The frequency spectrum of the gel was obtained between 0.1 Hz and 10 Hz at a strain of 5% and a temperature of 25°C.

7.2.5 *SCANNING ELECTRON MICROSCOPY (SEM)*

Scanning electron microscopy (SEM) analysis was performed on the AX gels that were formed. The samples were lyophilized before analysis at –40°C/0.133 mbar using a Freezone 6 freeze dryer (Labconco, Kansas City, USA). The analysis was conducted at the Microscopy Laboratory of the Department of Polymers at the University of Sonora, using a scanning electron microscope JEOL JSM-7401F (JEOL Ltd., Peabody, USA) operating at 1.8 kV in secondary electron mode.

7.2.6 *MTZ RELEASE KINETICS FROM THE AX GEL*

The gel had dimensions of 3 cm in diameter and 0.3 cm in height (sheet-type gel). A volume ratio of 1:3 was used, with 1 mL of gel per 3 mL of physiological solution as the recovery liquid. The experiments were conducted at different pH values (1.2, 4.5, and 6.8), 37°C, and 90 rpm. Every hour, 3 mL of sample was collected for a total of six hours. At each sampling time, the physiological solution was replaced. The collected samples were frozen for later MTZ content analysis (Krishnaiah et al., 2002).

7.2.7 *MTZ QUANTIFICATION*

The MTZ released from the AX gel was quantified using UV-Vis spectrophotometry, following the method described by Satyam et al. (2012).

A calibration curve was constructed using MTZ solutions with concentrations ranging from 0.1 to 0.3 mg/mL. The relationship between the MTZ concentration and the absorbance measured at 320 nm exhibited a linear correlation (99.85%), enabling us to utilize this curve for the quantification of MTZ released from the AX gels.

7.2.8 MTZ APPARENT DIFFUSION COEFFICIENT

The apparent diffusion coefficient (Dm) of MTZ during its release from the AX gel was determined by analyzing the release kinetics of the compound. Fick's second law of diffusion was applied to calculate the concentration of MTZ released as a function of time and distance (Crank, 1975).

$$\frac{Mt}{Mo} = \frac{4}{H}\left(\frac{Dmt}{\pi}\right)^{0.5}$$

where; 'Mt' is the cumulative mass of MTZ released at time 't;' 'Mo' is the amount of MTZ in the gel at time zero; 't' is the time, 'L' is the height of the gel (0.4 cm); and 'Dm' is the apparent diffusion coefficient of the molecule under study. The accumulated mass of MTZ released (Mt/Mo) vs. the square root of time is a linear relationship whose slope allows us to calculate Dm (Carvajal-Millan et al., 2005b). The percentage of MTZ released at the end of the experiment was also calculated.

7.2.9 STATISTICAL ANALYSIS

Measurements were performed in triplicate. All results are expressed as mean values.

7.3 RESULTS

7.3.1 FORMATION OF AX AND AX/MTZ GELS

The formation of gels at 5% AX (w/v) was studied using low-strain dynamic rheology. The AX and AX/MTZ solutions exhibited an increase in the elastic modulus (G') over time until reaching a pseudo plateau (Figures 7.1 and 7.2, respectively). The value of G' was equal to that of the viscous modulus (G”) with tan = (G”/G’) equal to one at 3.5 and 8.5 min for AX and AX/MTZ, respectively. These gel time (tg) values are lower than those reported in other wheat AX, where the tg varies between 10 and 25 min (Carvajal-Millan et al., 2005a; Berlanga-Reyes et al., 2014). This result could be explained based on the lower molecular weight of wheat bran AX used in the present study (74 kDa) compared to wheat endosperm AX (300–400 kDa). This higher molecular weight in wheat AX forms more viscous AX solutions, retarding the action of laccase, which is reflected in an initial phase of little increase in G,’ which is not observed in wheat bran AX used in the present study.

On the other hand, wheat bran AX has a lower ferulic acid content (0.02 μg/mg AX) compared to wheat endosperm AX (1–2 μg/mg AX), which could indicate that the initial stage of gelation of wheat bran AX is mainly determined by the shorter length of the chains rather than by the content of ferulic acid. However, more specific studies are required to elucidate the molecule’s gelation mechanism and explain this behavior in depth. At the end of the experiment, the G’ values were 175 and 164 Pa for the AX and AX/MTZ gels, respectively (Figures 7.1 and 7.2, respectively).

The mechanical spectra of the gels after 1.5 h at 25°C are shown in Figures 7.3 and 7.4, respectively. These spectra are typical of gels with linear and strain frequency-independent G’ values and much smaller, frequency-dependent G” values (Carvajal-Millan et al., 2005a). This behavior is similar to that previously reported for wheat AX gelled with laccase or the peroxidase/H_2O_2 system (Izydorczyk & Biliaderis, 1995; Dervilly et al., 2000; Vansteenkiste et al., 2004; Carvajal-Millan et al., 2005a). The values of tan = (G”/G’) calculated at 0.25 Hz were 0.10 and 0.11 for the AX and AX/MTZ gels, indicating the presence of a flexible system. The slight decrease in the tan delta value in the AX gel relative to the AX/MTZ gel indicates a lower viscous contribution to the gel structure.

7.3.2 CONTENT OF COVALENT BONDS IN THE AX GEL AND AX/MTZ GEL

At the end of gelation, the AX and AX/MTZ gels exhibited a di-FA content of 0.01 μg/mg AX, indicating that MTZ does not impact the formation of covalent cross-links in the AX gel. This di-FA content falls within the range reported for other wheat AX gels (Carvajal-Millan et al., 2005a, b; Berlanga-Reyes et al., 2014). The detected isomeric structures of di-FA in

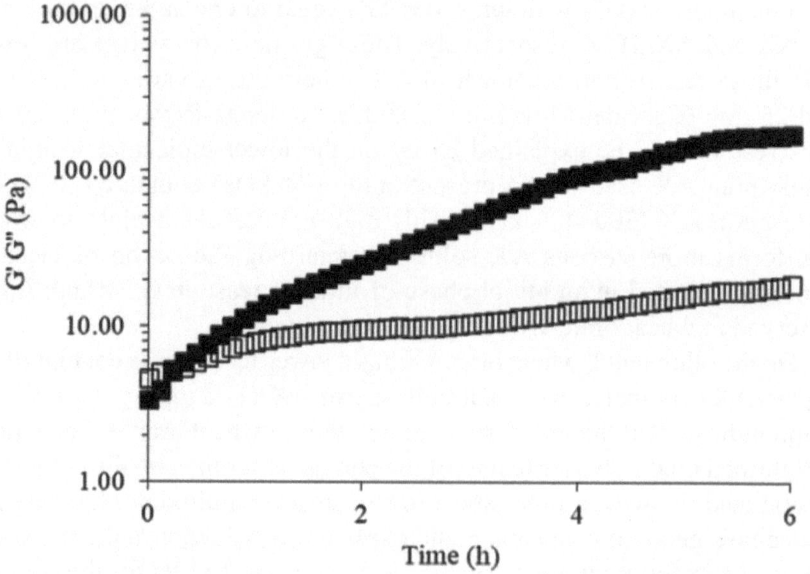

FIGURE 7.1 Changes in elasticity (G' □) and viscosity (G" ■) of an AX solution during laccase-induced gelation at 25°C, 0.25 Hz, and 5% strain.

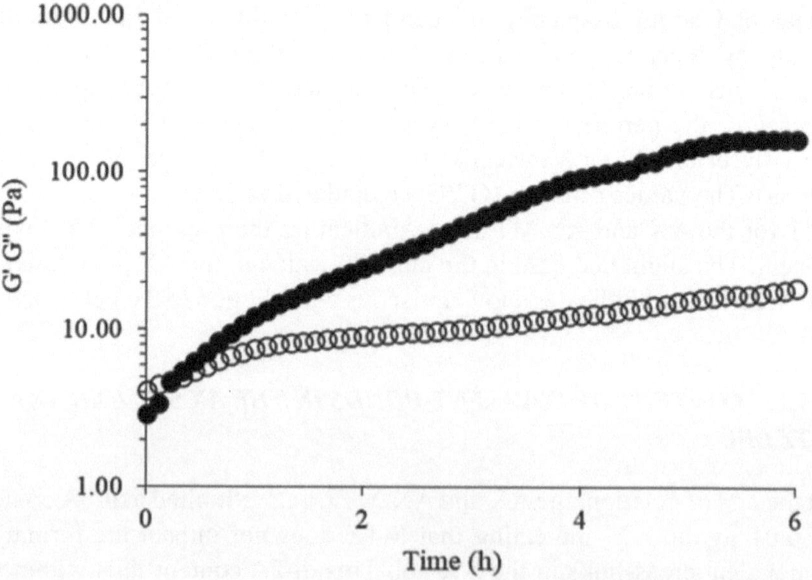

FIGURE 7.2 Changes in elasticity (G' ○) and viscosity (G" ●) of an AX/MTZ solution during laccase-induced gelation at 25°C, 0.25 Hz, and 5% strain.

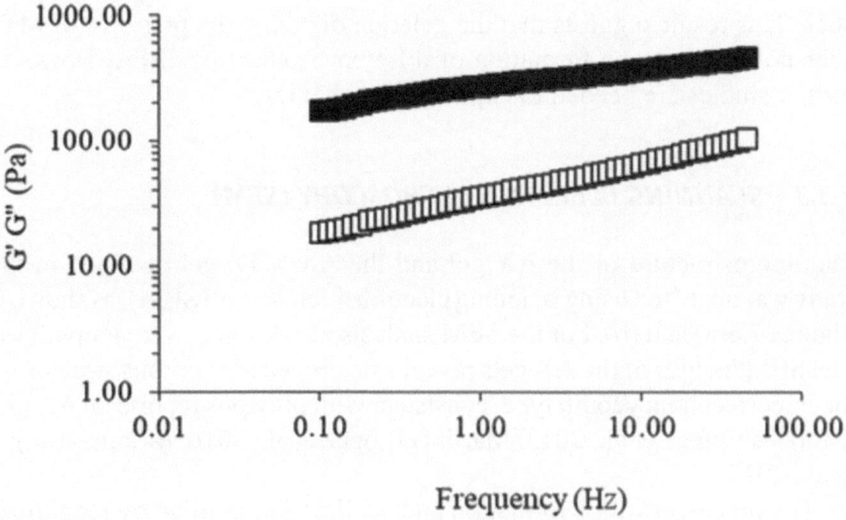

FIGURE 7.3 Frequency spectrum (G' □, G" ■) of AX gel after 1.5 h of gelation by laccase at 25°C. Measurement at 25°C and 5% deformation.

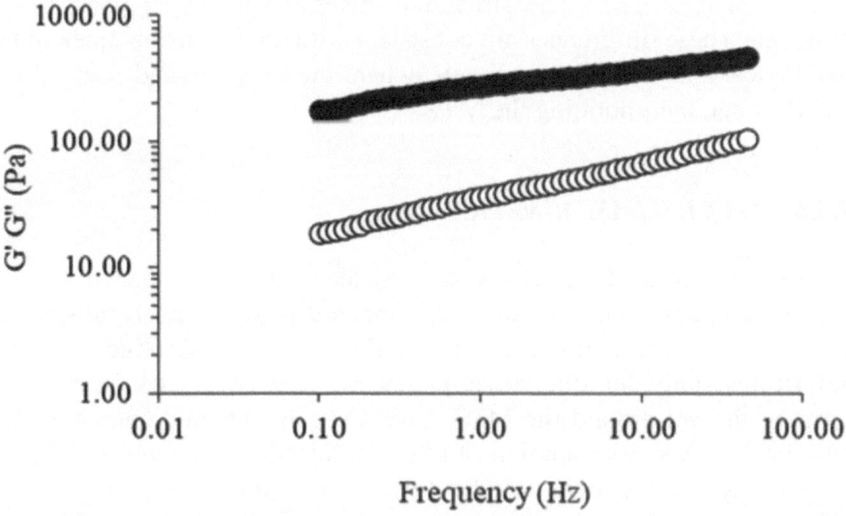

FIGURE 7.4 Frequency spectrum (G' ○, G" ●) of AX/MTZ gel after 1.5 h of gelation by laccase at 25°C. Measurement at 25°C and 5% deformation.

the AX and AX/MTZ gels were 5-5′ and 8-O-5′, both at a percentage of 50%. This result suggests that the gelation of AX in the presence of MTZ does not result in the formation of different isomers of di-FA. However, further studies are needed to explain this behavior.

7.3.3 SCANNING ELECTRON MICROSCOPY (SEM)

The microstructure of the AX gel and the AX/MTZ gel obtained in this study was analyzed using scanning electron microscopy (SEM), as shown in Figures 7.5(a) and (b). For the SEM analysis, the AX gels were lyophilized. The SEM images of the AX gels reveal a heterogeneous porous structure of the imperfect honeycomb type, consistent with previous reports on AX gels (Morales-Ortega et al., 2013; Martinez-Lopez et al., 2016; Morales-Burgos et al., 2017).

The processes of ice formation and sublimation implied by lyophilization are determinants of the material's morphology. Consequently, it is affected in various ways by the freezing conditions and other thermal phenomena that occur during the drying process. It can be observed that the AX gel has a structure with smaller and more continuous porosities. In the AX/MTZ gel, a less structured material with larger porosities is observed. These differences are consistent with the results obtained in the rheological analysis of the AX gels, where the G' decreased from 175 to 164 Pa when incorporating the MTZ.

7.3.4 MTZ RELEASE KINETICS

In most controlled drug release studies, the molecule is loaded into the gel through diffusion. However, this method is slow and results in the loss of some charged molecules since only a portion is introduced into the gel. In this study, an alternative method was employed, which involved forming the gel around the MTZ. The AX gels obtained contained 200 mg of MTZ. A similar amount of MTZ in 20% guar gum was previously reported (Krishnaiah et al., 2002). However, that study did not register the effect of MTZ on the viscoelastic properties of the formed material.

Figure 7.6 illustrates the kinetics of MTZ release from the 5% (w/v) AX gel. The cumulative mass (Mt/Mo) of released MTZ as a function of time increased from 0 to 0.18. Additionally, the relationship between

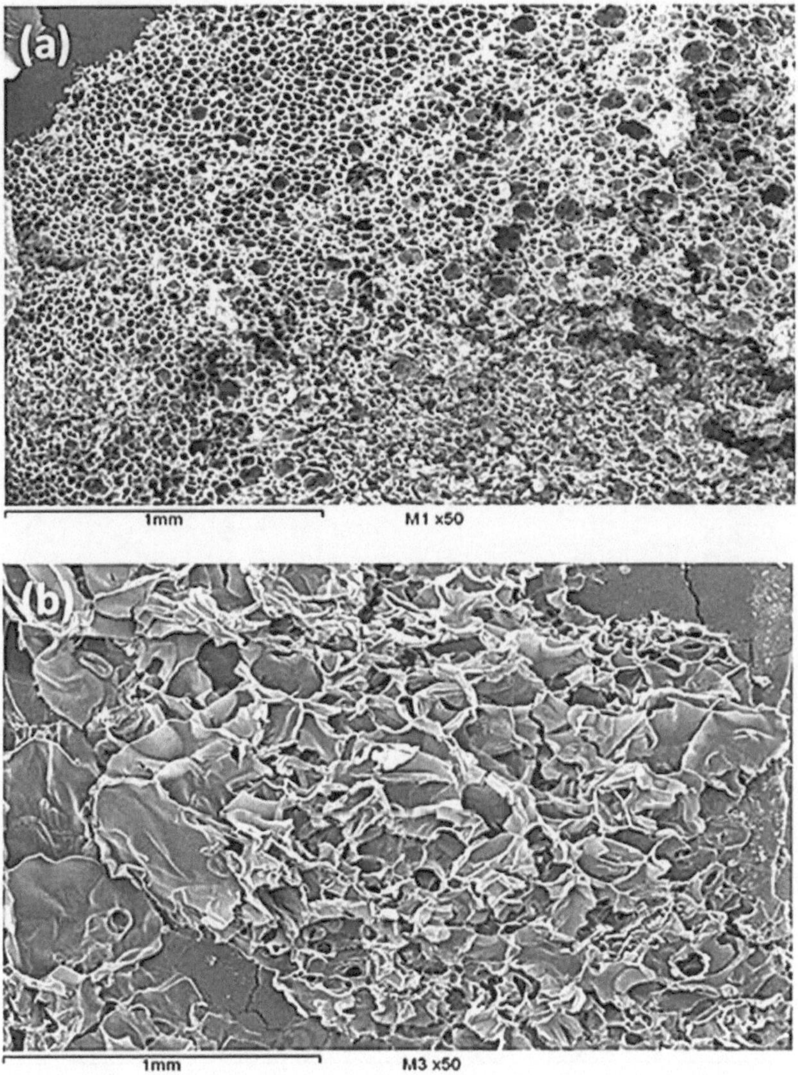

FIGURE 7.5 Surface microstructure of (a) AX gel; and (b) AX/MTZ gel analyzed by scanning electron microscopy (50x magnification).

the total amount of released MTZ (Mt/Mo) and the square root of time is shown, enabling the calculation of the apparent diffusion coefficient (Dm) of MTZ in the AX gels during the period, which exhibits a linear relationship. Table 7.1 presents the Dm value of MTZ found in the AX gel and the diffusion coefficient value of MTZ in water (Do) obtained from the literature (Meulemans et al., 1989). After 5 hours of incubation of the AX gels, only 18% of the initially incorporated total MTZ was released. The average time between food ingestion and its arrival in the colon is 6–9 hours. These results indicate that most of the MTZ incorporated into the AX gel would not be released during its passage through the upper gastro-intestinal system (first 5 hours). The AX in the colon could be degraded by the intestinal microflora (Martinez-Lopez et al., 2016), allowing the release of MTZ, which could exert its antiparasitic action in treating amoebiasis in the colon.

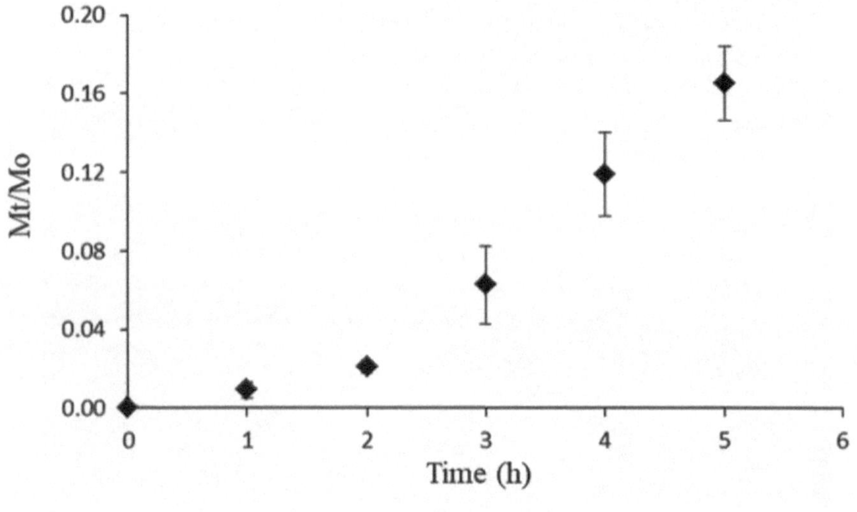

FIGURE 7.6 Metronidazole release kinetics from 5% (w/v) AX gels.

TABLE 7.1 Metronidazole Release Parameters from the 5% (w/v) AX Gel

D_0 *(× 10^{-6}) (cm^2/s)	D_m (× 10^{-7}) (cm^2/s)	MTZ Released (%)
6.4	3.95 ± 0.80	18 ± 2

**D_0 is the diffusion coefficient of metronidazole in water at 37°C reported in the literature (Meulemans et al., 1989).*

7.4 CONCLUSIONS

Wheat bran AX forms covalent gels in the presence of MTZ using laccase as a cross-linking agent. The MTZ trapped in the AX gels can be released in a controlled manner during the first 5 hours, allowing for the calculation of its apparent diffusion coefficient. The low rate of MTZ exit from the gel means that only 18% of the molecule is lost through this mechanism in the first 5 hours. This property enables the delivery of most of the drug loaded in the gel once the polymeric network is fermented in the colon by the microbiota. The ability of wheat bran AX to encapsulate MTZ offers an alternative use for wheat bran as a source of this high-value-added polysaccharide.

KEYWORDS

- **arabinoxylans**
- **controlled release**
- **covalent bonds**
- **cross-linking**
- **diffusion coefficient**
- **dimers**
- **ferulic acid**
- **gels**
- **trimers**
- **wheat by-product**

REFERENCES

Adams, L. E., Kroon, P. A., Williamson, G., & Morris, V. J., (2003). Characterization of heterogeneous arabinoxylans by direct imaging of individual molecules by atomic force microscopy. *Carbohydrate Res.*, *338*, 771–780. doi: 10.1016/s0008-6215(03)00017-x.

Andrewartha, K. A., Phillips, D. R., & Stone, B. A., (1979). Solution properties of wheat-flour arabinoxylans and enzimatically modified arabinaoxylans. *Carbohydrate Res.*, *77*, 191–204.

Autio, K., (2006). Functional aspects of cereal cell-wall polysaccharides. In: Eliasson, A. C., (ed.), *Carbohydrates in Food* (pp. 167–207). CRC-Taylor & Francis. Boca Raton, FL.

Bendesky, A., & Menéndez, D., (2001). Metronidazol: Una visión integral. *Rev. Fac. Med. UNAM, 44*(6), 255–259.

Berlanga-Reyes, C. M., Carvajal-Millan, E., Hicks, K. B., Yadav, M. P., Rascón-Chu, A., Lizardi-Mendoza, J., Toledo-Guillén, A. R., & Islas-Rubio, A. R., (2014). Protein/

arabinoxylans gels: Effect of mass ratio on the rheological, microstructural, and diffusional characteristics. *Int J Mol Sci., 15*(10), 19106–19118. https://doi.org/10.3390/ijms151019106.

Berlanga-Reyes, C. M., Carvajal-Millán, E., Juvera, G. C., Rascón-Chu, A., Márquez-Escalante, J. A., & Martínez-López, A. L., (2009). Laccase induced maize bran arabinoxylan gels: Structural and rheological properties. *Food Sci Biotech., 18*(4), 1027–1029.

Berlanga-Reyes, C. M., Carvajal-Millan, E., Lizardi-Mendoza, J., Islas-Rubio, A. R., & Rascón-Chu, A., (2011). Enzymatic cross-linking of alkali extracted arabinoxylans: Gel rheological and structural characteristics. *Int J Mol Sci., 12*(9), 5853–5861. https://doi.org/10.3390/ijms12095853.

Borro, M. F., (2001). *Estudio de la Eficacia de los Aglutinantes en la Disolución del Metronidazol Tabletas 250 mg*. Doctoral thesis. Universidad de Guayaquil. Guayaquil, Ecuador.

Carvajal-Millan, E., Guigliarelli, B., Belle, V., Rouau, X., & Micard, V., (2005a). Storage stability of arabinoxylan gels. *Carbohydr Polym., 59*(2), 181–188. https://doi.org/10.1016/j.carbpol.2004.09.008.

Carvajal-Millan, E., Guilbert, S., Doublier, J. L., & Micard, V., (2006). Arabinoxylan/protein gels: Structural, rheological, and controlled release properties. *Food Hydrocolloids, 20*(1), 53–61. https://doi.org/10.1016/j.foodhyd.2005.02.011.

Carvajal-Millan, E., Landillon, V., Morel, M. H., Rouau, X., Doublier, J. L., & Micard, V., (2005b). Arabinoxylan gels: Impact of the feruloylation degree on their structure and properties. *Biomacromolecules, 6*(1), 309–317. https://doi.org/10.1021/bm049629a.

Carvajal-Millan, E., Rascón-Chu, A., Márquez, E. J., Micard, V., Ponce De, L. N., & Gardea, A., (2007). Maize bran gum: Extraction, characterization, and functional properties. *Carbohydr. Polym., 69*(2), 280–285. https://doi.org/10.1016/j.carbpol.2006.10.006.

Crank, J., (1975). *The Mathematics of Diffusion* (2nd edn., p. 414). Oxford Press.

Dervilly, G., Saulnier, L., Roger, P., & Thibault, J. F., (2000). Isolation of homogeneous fractions of wheat water-soluble arabinoxylan. Influence of structure on their macromolecular characteristics. *J Agric Food Chem., 48*(2), 270–278. https://doi.org/10.1021/jf990222k.

Figueroa-Espinoza, M. C., & Rouau, X., (1998). Oxidative cross-linking of pentosans by a fungal laccase and horseradish peroxidase: Mechanism of linkage between feruloylates arabinoxylans. *Cereal Chem., 75*(2), 259–265. https://doi.org/10.1094/CCHEM.1998.75.2.259.

Grootaert, C., Delcour, J. A., Courtin, C. M., Broekaert, W. F., Verstraete, W., & Van De, W. T., (2007). Microbial metabolism and prebiotic potency of arabinoxylan oligosaccharides in the human intestine. *Trends Food Sci Technol., 18*(2), 64–71. https://doi.org/10.1016/j.tifs.2006.08.004.

Halliwell, B., (1978). Lignin synthesis: The generation of hydrogen peroxide and superoxide by horseradish peroxidase and its stimulation by manganese (II) and phenols. *Planta, 140*, 81–88. https://doi.org/10.1007/BF00389384.

Hatfield, R. D., Ralph, J., & Grabber, J. H., (1999). Cell wall cross-linking by ferulates and diferulates in grasses. *J Sci Food Agric., 79*(3), 403–407. https://doi.org/10.1002/(SICI)1097-0010(19990301)79:3<403::AID-JSFA263>3.0.CO;2-.

Ishii, T., (1997). Structure and functions of feruloylated polysaccharides. *Plant Sci., 127*(2), 111–127. https://doi.org/10.1016/S0168-9452(97)00130-1.

Izydorczyk, M. S., & Biliaderis, C. G., (1995). Cereal arabinoxylan: Advances in structure ad physicochemical properties. *Carbohydr Polym.*, *28*(1), 33–48. https://doi.org/10.1016/0144-8617(95)00077-1.

Izydorczyk, M. S., & Biliaderis, C. G., (2007). Arabinoxylans: Technologically and nutritionally functional plant polysaccharides. In: Biliaderis, C. G., & Izydorczyk, M. S., (eds.), *Functional Food Carbohydrates* (pp. 249–290). CRC Press, Boca Raton. FL, USA.

Izydorczyk, M. S., Biliaderis, C. G., & Bushuk, W., (1991). Comparison of structure and composition of water-soluble pentosans from different wheat varieties. *Cereal Chem.*, *68*(2), 139–144.

Krishnaiah, Y. S., Bhaskar, R. P. R., Satyanarayana, V., & Karthikeyan, R. S., (2002). Studies on the development of oral colon targeted drug delivery systems for metronidazole in the treatment of amoebiasis. *Int. J. Pharm.*, *236*(1, 2), 43–55. https://doi.org/10.1016/S0378-5173(02)00006-6.

Lapierre, C., Pollet, B., Ralet, M. C., & Saulnier, L., (2001). The phenolic fraction of maize bran: Evidence for lignin-heteroxylan association. *Phytochemistry, 57*(5), 765–772. https://doi.org/10.1016/S0031-9422(01)00104-2.

Lau, A. H., Lam, N. P., & Piscitelli, S. C., (1992). Clinical pharmakokinetics of metronidazole and other nitroimidazole anti-infectives. *Clin. Pharmacokinet.*, *23*(5), 328–364. doi: 10.2165/00003088-199223050-00002.

Martinez-Lopez, A. L., Carvajal-Millan, E., Micard, V., Rascón-Chu, A., Brown-Bojorquez, F., Sotelo-Cruz, N., López-Franco, Y. L., & Lizardi-Mendoza, J., (2016). *In vitro* degradation of covalently cross-linked arabinoxylan hydrogels by bifidobacteria. *Carbohydr. Polym.*, *144*, 76–82. https://doi.org/10.1016/j.carbpol.2016.02.031.

Meulemans, A., Paycha, F., Hannoun, P., & Vulpillat, M., (1989). Measurement and clinical and pharmacokinetic implications of diffusion coefficients of antibiotics in tissues. *Antimicrob Agents Chemother.*, *33*(8), 1286–1290. https://doi.org/10.1128/AAC. 33.8.1286.

Morales-Burgos, A. M., Carvajal-Millan, E., Lopez-Franco, Y. L., Rascon-Chu, A., Lizardi-Mendoza, J., Sotelo-Cruz, N., Brown-Bojorquez, F., et al., (2017). Syneresis in gels of highly ferulated arabinoxylans: Characterization of covalent cross-linking, rheology, and microstructure. *Polymers, 9*(5), 164; https://doi.org/10.3390/polym9050164.

Morales-Ortega, A., Niño-Medina, G., Carvajal-Millán, E., Gardea-Béjar, A., Torres-Chávez, P., López-Franco, Y., Rascón-Chu, A., & Lizardi-Mendoza, J., (2013). Los arabinoxilanos ferulados de cereales. Una revisión de sus características fisicoquímicas y capacidad gelificante. *Rev. Fitotec. Mex.*, *36*, 439–446.

Niño-Medina, G., Carvajal-Millán, E., Lizardi, J., Rascon-Chu, A., Marquez-Escalante, J. A., Gardea, A., Martinez-Lopez, A. L., & Guerrero, V., (2009). Maize processing waste water arabinoxylans: Gelling capability and cross-linking content. *Food Chem.*, *115*(4), 1286–1290. https://doi.org/10.1016/j.foodchem.2009.01.046.

Niño-Medina, G., Carvajal-Millán, E., Rascon-Chu, A., Márquez-Escalante, J. A., Guerrero, V., & Salas-Muñoz, E., (2010). Feruloylated arabinoxylans and arabinoxylan gels: Structure, sources, and applications. *Phytochem Rev.*, *9*(1), 111–120. https://doi.org/10.1007/s11101-009-9147-3.

Papon, P., Leblond, J., & Meijer, P., (2006). Gelation and transitions in biopolymers. In: *The Physics in Phase Transition. Concepts and Applications* (2nd edn., pp. 189–213). Springer-Verlag. New York.

Ralph, E. D., (1983). Clinical pharmacokinetics of metronidazole. *Clin Pharmacokinet.*, *8*(1), 43–62. https://doi.org/10.2165/00003088-198308010-00003.

Rattan, O., Izydorczyck, M. S., & Biliaderis, C. G., (1994). Structure and rheological behavior of arabinoxylans from Canadian bread wheat flours. *Lebensm Wiss Technol.*, *27*(6), 550–555. https://doi.org/10.1006/fstl.1994.1108.

Roubroeks, J. P., Andersson, R., Mastromauro, D. I., Christensen, B. E., & Åman, P., (2001). Molecular weight, structure and shape of oat (1→3),(1→4)-b-D-glucan fractions obtained by enzymatic degradation with (1→4)-b-D-glucan 4-glucanohydrolase from *Trichoderma reesei*. *Carbohydr Polym.*, *46*(3), 275–285. https://doi.org/10.1016/S0144-8617(00)00329-5.

Satyam, Z., Dongare, U. S., Jadhav, S. A., & Jadhav, M. B., (2012). Validated spectrophotometric methods for simultaneous estimation of metronidazole and furazolidone in pure and tablet dosage form. *Int Res J Pharm.*, *4*(3), 4805–4813.

Saulnier, L., Sado, P. E., Branlard, G., Charmet, G., & Guilon, F., (2007). Wheat arabinoxylan: Exploiting variation in amount and composition to develop enhanced varieties. *J. Cereal Chem.*, *46*(3), 261–281. https://doi.org/10.1016/j.jcs.2007.06.014.

Schwartz, D. E., & Jeunet, F., (1976). Comparative pharmacokinetics studies of ornidazol and metronidazole in man. *Chemotherapy*, *22*(1), 19–29. doi: 10.1159/000221906.

Selinheimo, E., Autio, K., Kruus, K., & Buchert, J., (2007). Elucidating the mechanism of laccase and tyrosinase in wheat bread making. *J. Agric. Food Chem.*, *55*(15), 6357–6365. doi: 10.1021/jf0703349.

Smith, M. M., & Hartley, R. D., (1983). Occurrence and nature of ferulic acid substitution of cell-wall polysaccharides in graminaceous plants. *Carbohydr. Res.*, *118*, 65–80. https://doi.org/10.1016/0008-6215(83)88036-7.

Vansteenkiste, E., Babot, C., Rouau, X., & Micard, V., (2004). Oxidative gelation of feruloylated arabinoxylan as affected by protein. Influence on protein enzymatic hydrolysis. *Food Hydrocoll.*, *18*, 557–564. https://doi.org/10.1016/j.foodhyd.2003.09.004.

CHAPTER 8

Performance Evaluation for Sustainable Urban Water Supply Services in Mexico: A Case Study

ISMAEL AGUILAR-BENITEZ[1] and SERGIO COLIN CASTILLO[2]

[1]*EL Colegio de la Frontera Norte (COLEF), Sede Monterrey, Mexico*

[2]*Centro de Investigaciones Socioeconómicas, Universidad Autónoma de Coahuila, Mexico*

ABSTRACT

The efficiency of water service providers is analyzed by applying robust non-parametric techniques, using data generated by the Economic Censuses in Mexico. Five efficiency estimates were obtained: (i) data envelope analysis (DEA); (ii) free disposal hull (FDH); (iii) order-*m* estimator; (iv) order-alpha estimator; and (v) biased corrected (BC) estimator. This work has two main contributions: firstly, a microdata census database is used for the first time to estimate the technical efficiency of water utilities in Mexico; and secondly, a comparative analysis of the effectiveness of these estimators is performed in a context of a strong presence of outliers and errors in the data. The results allow for the identification and discussion of efficiency measures for each water utility, the influence of possible error margins, and recommendations for improving data collection, especially for supervising utilities that report inaccurate data.

Sustainable Environment and Health: Practical Strategies. Leonardo Sepúlveda Torre, Ann Rose Abraham, Neha Kanwar Rawat, & A. K. Haghi (Eds.)

8.1 INTRODUCTION

The evaluation of the efficiency of businesses (utilities) that provide drinking water, drainage, and sanitation services delivers relevant information for decision-makers and the design of public policies (Picazzo-Tadeo et al., 2009). Water-efficient service is a necessary condition to achieve quantity, quality, and continuity, which are desirable characteristics to protect the human right to water, sanitation, and health. Lockdowns, such as the recent health contingency due to the COVID-19 coronavirus, have tested the utilities and have highlighted the importance of having water with such desirable features for household use. In practice, both in Mexico and other countries, water and sanitation services operate as monopolies due to special characteristics such as significant investments and large economies of scale. Most of the time, government subsidies are necessary for their operation (Griffin, 2016; Aguilar Benitez, 2011). Certainly, most water utilities in Mexico are operated under public administration and are financially dependent on public subsidies. Both characteristics, monopoly, and dependence on public subsidies, contribute to the lack of incentives to achieve high-efficiency levels.

From an economic perspective, technical efficiency is an aggregated measure that expresses the ratio of results (outputs) to a set of factors (inputs). Since there is no unique production function to define an absolute technological standard of achievable optimum given a set of inputs, a viable option is to measure the relative efficiency of each unit within their respective group, population, or sample. For empirical analysis, two main approaches have been proposed: a parametric technique based on econometric methods, known as stochastic frontier analysis (SFA), and a non-parametric technique using mathematical programming. The validity of any stochastic production frontier depends on the relevance and suitability of the functional form of the model (e.g., Cobb-Douglas). An advantage of non-parametric methods is that they do not require assuming a technical functional form of the model to define the relationship between inputs and outputs; in other words, the production function is not specified.

This work analyzes the efficiency of water service providers in Mexico applying robust non-parametric techniques to microdata generated through the questionnaire responded by each organism that carry out the collection, treatment, and supply of water (OCTSA) in Mexico[1]. The data used is part of the Economic Census in Mexico[2] and, to our knowledge, it has not been used for a similar purpose before. Indeed, the only reference of a non-parametric analysis of water services in Mexico, applying DEA, assesses the decentralization effect of 110 operating agencies (Anwandter & Ozuna, 2002). However, in the international arena, the application of the non-parametric approach to estimate efficiency frontiers for the provision of water services has been extensive in countries such as the United Kingdom, the United States, Australia, Spain, and to a lesser extent in Latin America (e.g., Brazil, *see* Worthington, 2014). Recently, issues such as economies of scale, outliers, data validation, and statistical "noise" have been discussed in the context of water services, by applying non-parametric methods (*see* Güngör-Demirci et al., 2018; Cabrera et al., 2018; Molinos-Senante et al., 2016; Ananda, 2014; De Witte & Marques, 2010; Sampaio de Souza & Stosic, 2005).

We compare five efficiency estimates of water service providers in Mexico. These estimations were derived using five nonparametric techniques:

1. **Data Envelopment Analysis (DEA):** This assuming strict convexity on the set of production possibilities given a set of units, i.e., firms (Charnes et al., 1978).
2. **Free Disposal Hull (FDH):** This relaxing the assumption of convexity required by DEA, but assuming the full availability in the observed enveloped curve (Deprins, Simar, & Tulkens, 1984).
3. **Order-m Estimator:** This comparing each unit in a random subsample of observations (Cazals, Florens, & Simar, 2002).
4. **Order-Alpha Method (α-Order):** This which estimates efficiency based on conditional quantiles of units being compared (Aragón et al., 2005).
5. **The Biased Corrected (BC) Estimator (Kneip et al., 2008):** This using bootstrap technique to replicate random samples and

[1] This work is a result of the project "Towards an Integrated Water Management by Hydrological Basin: An Analysis of Availability and Uses" Project funded by Conacyt in the 2014 National Problems Call, No.: 248719.

[2] The collection, treatment and supply of water is reported under the classification 2221, according to the North America Industrial Classification System (SCIAN, 2013). The use of microdata for this analysis was carried out at the INEGI Microdata Laboratory, and it was registered under the project LM588. The author appreciates the support of the Microdata Laboratory staff.

compare each unit[3]. These estimators are described in the methodology section.

The main contributions of this work are twofold. First, from an empirical standpoint, it is the first time that the microdata database from the Economic Census is used to estimate the technical efficiency of the Organisms that carry out the Collection, Treatment, and Supply of Water (OCTSAs) in Mexico. We demonstrate its potential for conducting this analysis. Second, from a methodological perspective, it aims to contribute to the comparative analysis of the effectiveness of technical efficiency estimators, considering a significant incidence of outliers and errors in the data. Ultimately, we aim to provide relevant information that can be useful for decision-makers. The measurement of efficiency and the identification of causes for each unit are valuable pieces of information derived from economic analysis, and hopefully, they can serve as clues for management improvement.

The writing is organized as follows: Section 8.2 presents the characteristics of the OCTSAs in Mexico in 2014. Section 8.3 describes the nonparametric estimators applied in this work. In Section 8.4, we present the main results and discuss the efficacy of the five estimators in reducing bias caused by a high incidence of outliers. Finally, we conclude with some key findings in Section 8.5.

8.2 CHARACTERISTICS OF THE ORGANISMS THAT CARRY OUT THE COLLECTION, TREATMENT, AND SUPPLY OF WATER (OCTSAS) IN MEXICO (2014)

A total of 2,688 OCTSAs were registered in Mexico in 2014. An OCTSA can provide services to any given municipality area, and in some cases, also supply water services to several municipalities or even one State (e.g., Monterrey OCTSA supplies the State of Nuevo León). Out of the total, only 1,394 OCTSAs offer services exclusively to urban areas, while 1,294 serve both urban and rural populations.

According to Figure 8.1, 33% of the OCTSAs belong to communities with less than 2,500 inhabitants (rural areas), while the most frequent

[3] For a review of the works with the greatest influence on productivity and efficiency analysis using the nonparametric approach, see the book by Daraio & Simar (2007).

size falls within the range of 2,501 to 20,000 inhabitants (48%). Only 10 communities have a population larger than 1 million. In 2013, the OCTSAs employed 1,22,798 people, with 96% hired directly by the company and 4% being subleased labor from another source. The size of OCTSAs can vary, with some being very small, having only a few employees and serving a limited number of connections, while others grow beyond their optimal size in large cities. The organization of OCTSAs is quite diverse, functioning as operating agencies, water systems, local boards, municipal committees, and private water concessionaires. Out of the total, 2,401 OCTSAs (90%) are classified as public sector services, 257 (9%) are registered as civil associations, and only 30 (1%) operate as commercial companies or cooperative societies.

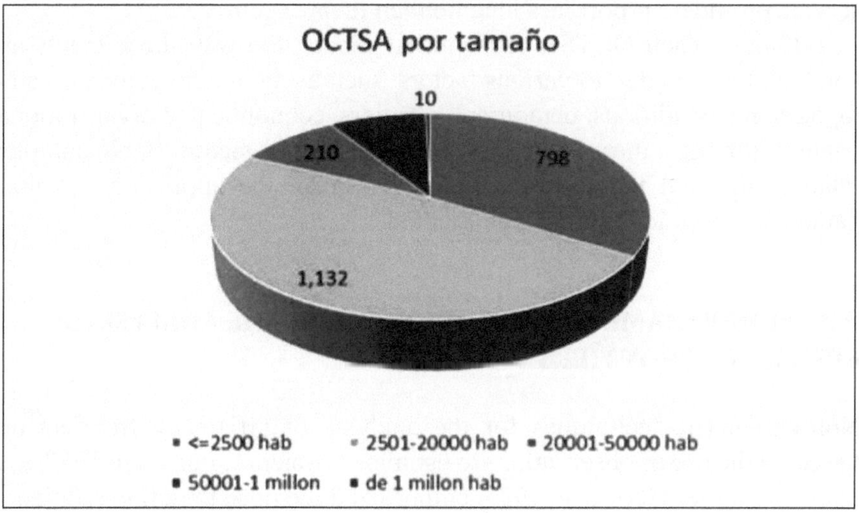

FIGURE 8.1 OCTSA by size in Mexico (2014).

Source: Own elaboration using 2014 Economic Censuses data.

The total water managed by the OCTSAs was 12,400.6 million cubic meters per year, considering the extraction from surface sources (27%), underground sources (61%), and block water (12%). Due mainly to the depletion or overexploitation of the sources, several OCTSAs require purchasing bulk water to supplement the volume of water they need.

According to the reports, only half (51%) of the supplied water is charged. This is a crucial point for the management because the expenses to supply water services in Mexico were reported for a total of 24,935.3 million pesos. On average, the main expenses were energy (39.5%); bulk water supply (16.1%); repairs and maintenance (9.8%); physical-chemical agents and other materials for the water distribution service (7.6%), and fuels, lubricants, and oil (5.4%). The remaining (21.6%) paid other expenses such as advertising, office, communication, professional services, insurance premiums, among others.

The income from the supply of goods and services for the OCTSAs was 47,561.2 million pesos. This revenue primarily came from water supply (69.3%), connection, and sewerage rights (11.8%), income from previous years (9.2%), sale of treated water (2.1%), and other items (7.6%). The latter item includes the sale of sludge, advertising revenue, and the distribution of water through tank trucks. In the end, 35% of water service providers report less income than their expenses.

Although each OCTSA uses similar inputs, the way these inputs are combined differs due to various factors, such as the main source of water, topographic conditions, demographic factors, economic and organizational context, among others. The presence of multiple factors, their complex relationship, and the resulting outcomes make the application of non-parametric techniques a relevant method.

8.3 NON-PARAMETRIC TECHNIQUES FOR THE ANALYSIS OF EFFICIENCY FRONTIERS

Non-parametric techniques for the analysis of efficiency frontiers are based on the use of observations to estimate efficiency. Farrell in 1957, and later Charnes & Cooper in 1985, emphasized the need to define efficiency (technical or economic) by comparing observed best practices with a reference set or comparison group (Daraio & Simar, 2007). According to this concept, the frontier represents the upper surface of the production function, which is derived from the input-output combinations that yield the maximum achievable output given a production technology.

To estimate such boundaries, the non-parametric approach requires two conditions: orientation and returns to scale. The first condition is the inputs-oriented model, which emerges when decision-making units (DMUs) aim to achieve the maximum proportional reduction of inputs, given an output level, while staying within the production possibilities frontier. Conversely, the output-oriented model occurs when DMUs seek the maximum proportional increase in outputs, given the level of inputs, while remaining within the production possibilities frontier. The second condition is the returns to scale, which will be constant (CCR) if both the use of factors (inputs) and the output change in a similar proportion. It will be increasing if the output rises (or decreases) in a greater proportion than the factors increased, or it will be decreasing if the output varies in a smaller proportion than the factors. Non-parametric efficiency estimators are developed based on this general approach.

8.3.1 DATA ENVELOPMENT ANALYSIS (DEA) ESTIMATOR

The first non-parametric estimator of efficiency implemented for empirical analysis was the DEA, which compared DMUs considering all inputs and possible outputs (Cooper et al., 2011). The use of mathematical programming for DEA empirical application, using constant returns to scale (i.e., CCR), was modeled by Charnes, Cooper, & Rhodes (1978). The variable returns to scale (i.e., BCC) for DEA were developed by Banker, Charnes, & Cooper (1984) by introducing the assumption of variable returns to scale (VRS). This is, as we will see, the most appropriate assumption to use for this study.

The BCC linear programming for DEA model can be stated as follows: Suppose that you have n number of DMUs, and each DMUj (for $j = 1, 2, \ldots, n$), produce the same output but in different quantities, Yrj (for $r = 1, 2, \ldots, s$); using similar inputs also in different quantities Xij (for $i = 1, 2, \ldots, m$). Formally, the input-oriented model, assuming BCC, is formulated as follows (Bowlin, 1998):

Minimize: $\theta - \in [\sum_{i=1}^{m} s_i^- + \sum_{r=1}^{s} s_r^+]$

Subject to:

$$0 = \theta x_{i0} - \sum_{j=1}^{n} x_{ij}\lambda_j - s_i^-$$

$$y_{r0} = \sum_{j=1}^{n} y_{ij}\lambda_j - s_i^+$$

$$1 = \sum \lambda_j$$

$$0 \le \lambda_j;\ s_i^-,\ s_i^+\ for\, i = 1,\ldots,m;\ r = 1,\ldots,s;\ j = 1,\ldots,n$$

Graphically, the set of possible solutions creates an upper limit that surrounds all the observations, hence the name of enveloping analysis. The units on the frontier are efficient, and efficiency is defined as the radial distance from the real operation level to its corresponding point on the border (Farell, 1957). That efficiency is formally stated as $\theta(x_0, y_0) \leq 1$ to express the radial contraction (decrease in inputs) that the DMUi must achieve to be efficient. DEA assumes strict convexity on the set of production possibilities, e.g., all combinations are feasible, on a given set of theoretical possibilities.

8.3.2 FREE DISPOSAL HULL (FDH) ESTIMATOR

The second estimator, the free disposal hull (FDH), arises from the same principles of DEA but relaxes the convexity assumption. It assumes free availability of the production possibilities given the observed combinations or border points (assuming no error in the measurement). It reduces the possibilities to the empirically determined area by using the observed data (Deprins et al., 1984). In other words, the production set assumes a strong availability of outputs and free availability of inputs, creating a polyhedral envelope, but not necessarily convex. In linear programming, the input-oriented FDH estimator is formally stated as follows (Marques & De Witte, 2011):

$$\theta(\mathrm{x}, \mathrm{y}) = \min \{\theta \mid \theta \mathrm{x} \geq \sum_{i=1}^{n} \lambda_i x_i ; y \leq \sum_{i=1}^{n} \lambda_i y_i ; \sum_{i=1}^{n} \lambda_i = 1; \lambda_i \in \{0,1\} ; 1,\dots,n\}$$

These two estimators (DEA and FDH) are classified as deterministic because of their assumptions about determining the theoretical or observable production possibilities, given a data set without measurement errors. The main limitations of these estimators are their high sensitivity to outliers and the presence of errors in the data (Cazals et al., 2002).

8.3.3 M-ORDER ESTIMATOR

Cazals et al. (2002) start from the concept of expected minimum inputs (from an input-oriented model) to propose a third estimator: The M-order. Unlike the DEA and FDH, this estimator is based on partial boundaries by not including all data. In the simplest case (all DMUs produce a single unit), the M-order frontiers can be entered as follows: Considering a fixed integer $m \geq 1$. The lower limit of X, of order m, is defined as the minimum expected value of m random variables $x^1, x^2, \dots, x^m$, given a distribution function of X. The formal definition is offered by Daraio & Simar (2007, p. 68).

$$Øm = [min(x^1,\ldots,x^m) = \int_0^\infty [1-F_X(x)]^m\,dx\,]$$

In an input-oriented model, the symbol φm represents the expected minimum level of input among *m* DMUs. These DMUs are extracted from a set of companies that, in the simplest case, produce a single unit of output. The procedure for estimating the m-order efficiency can be summarized in four stages: (i) obtain a random sample of m DMUs, with replacement, from the group of DMUs that produce at least as much output as DMUi; (ii) calculate the partial efficiency using the obtained sample; (iii) repeat Steps 1 and 2 D times; and (iv) calculate the m-order efficiency as the average of the estimates obtained in Step 3. Due to the random resampling with replacement, a DMUi may or may not be available as its pair. As a result, input-oriented m-order estimators may exceed the value of 1, indicating super-efficiency beyond the estimated production frontier. However, one advantage of this estimator is that it allows the identification of units that can be classified as outliers, either because they are outside the estimated boundary or due to errors in the data.

In general, the expected frontier of order-m, φm (y), can be interpreted as the minimum expected value of achievable input among a fixed number (m) of DMUs. These DMUs are extracted from the population that produces at least one level of output Y. Finally, since the estimator of order m does not include all the observed data points, it serves as a benchmarking tool for a specific number of DMUs.

8.3.4 ALPHA ORDER ESTIMATOR (A-ORDER)

Similar to the order-m, the order-α estimator (or quantile-type boundaries estimator) assesses the boundary by setting the probability (1–α) of detecting points above this α-order boundary. In the simplest output case, the unity, the reference point will be the order-α quantile boundary, defined as the input level not exceeded by (1–α) × 100 firms that produce at least the same level of Y (output). Formally:

$$Ø(y) = inf\,\{x \mid F_{X|Y}(x|y) > 1-\alpha\}$$

The α-order estimator is similar to a "continuous version" of the M-order estimator, but it is based on the concept that for each company, there exists a quantile efficiency frontier within a set of comparison observations, where the company is efficient (either in terms of input or output). Therefore, if the quantile at which the company is efficient is $\alpha = 0.2$, it means that 80% (1–0.2 = 0.8) of DMUs in the comparison set outperform the company while using fewer inputs. When $\alpha = 100$, the estimator aligns with FDH, and if $\alpha < 100$, some DMUs can be classified as super-efficient and will not be encompassed by the estimated production frontier. Hence, similar to the m-order estimator, the α-order estimator can identify if a DMU is an outlier.

8.3.5 BC ESTIMATOR (BIASED CORRECTED ESTIMATOR)

Until this point, the estimators derived from the envelope analysis (DEA, FDH, and partial boundaries) measure the relative efficiency compared to an estimated frontier, using an unobservable generation process (Simar & Wilson, 2002). This limitation pertains to the knowledge of the statistical properties of these estimators. Kneip, Simar, & Wilson (2008) propose two bootstrap processes to approximate the asymptotic distribution of DEA estimators, considering VRS and an arbitrary number of inputs and outputs. These bootstraps are based on the idea of repeatedly simulating the data generation process by resampling and applying the original estimator to each simulated sample, so that the resulting estimates mimic the sampling distribution of that estimator. A key assumption for the data generating process is that the n observations in the sample (i.e., efficiency estimates) are independent and identically distributed (i.i.d) random variables in a possible convex set. Kneip et al. (2008) describe two algorithms in the following steps:

- **Algorithm I:** A data simulation is performed to extract n samples in each replica, creating pseudo samples for obtaining efficiency estimators. (i) A sample of m simulated data (bootstraps) is randomly generated from the original data set using uniform, independent sampling with replacement. (ii) Usual estimators of DEA efficiency (Farell coefficient) are calculated. (iii) Steps 1 and 2 are repeated B times to approximate the unknown distribution using the values obtained from the data simulation. And (iv) an asymmetric confidence interval is calculated for the estimators.

- **Algorithm II:** The second simulation approach is based on a "smoothing" process of the empirical data distribution and the production frontier, following the steps below:
    - o **Step 1:** A smoothed production frontier is calculated analogously to the observed one.
    - o **Step 2:** A series of bootstrap samples is extracted from the estimated smoothed density function, and the sample size is defined to estimate the efficiency.
    - o **Step 3:** DEA estimators are calculated for the sample size from Step 2.
    - o **Step 4:** The previous steps (Steps 2 and 3) are repeated, and the bootstrap values are used to approximate the unknown distribution function of the efficiency estimators.
    - o **Step 5:** The symmetric confidence interval is estimated.

These five estimators were applied to the OCTSA (2014) database generated in Mexico, as detailed in the next section.

8.3.6 PREPARATION OF THE DATABASE AND MODEL FOR THE NON-PARAMETRIC ANALYSIS

The original database includes information from 2,668 OCTSAs captured by the 2014 Economic Census in Mexico. The application of non-parametric methods requires non-negative values, therefore only strictly positive values are used. Around 188 missing values of drinking water distribution networks were eliminated, as well as the zero connections (69 OCTSA) and zero income (45 OCTSA). The inconsistent records of connections for domestic use were also identified and eliminated, specifically those with a number of industrial and commercial connections greater than domestic (35 observations). Finally, inconsistent extreme values were detected and eliminated; these were remarkably high in length network but did not correspond to the number of connections and personnel (two observations). In conclusion, the number of OCTSAs with complete information to calculate the non-parametric estimators was 2,349.

For the empirical model, the variables included were based on the literature review (*see* Ananda, 2014; Cabrera et al., 2018; De Witte & Marques, 2010; Güngör-Demirci et al., 2018; Marques & De Witte, 2011; Molinos-Senante et al., 2016; Picazo-Tadeo, 2009; Worthington, 2013), all of which have been previously used to analyze the activity of water services, and were chosen based on data availability. Four explanatory or input variables were chosen: (i) the length of networks; (ii) expenses on goods and services; (iii) the total personnel or staff; and (iv) the collected water. For the length of networks, the extensions (in km) of pipes of primary and secondary networks were added. The expenses (in current pesos 2013) on goods and services include items like physical-chemical inputs and supplies, electrical energy, and water supply in bulk. The variable "total personnel" is the number of technical and administrative staff under the direction and control of the OCTSA and those depending on another company. The collected water (in thousands of m^3) sums the volumes of surface and underground water obtained by each OCTSA, excluding block water.

For the output or dependent variable, the model uses the number of domestic connections. By using this variable instead of the income as the output, we recognize that OCTSAs in Mexico are non-profit organizations to supply piped water, drainage, and sanitation services. It is not realistic to see that OCTSA seeks profit. Indeed, about 35% of OCTSAs generate less income than expenses, restricting the use of income as an output variable. Table 8.1 describes the model variables in a wide range of values, e.g., the OCTSAs collect from 1,000 to more than a billion m^3 of water and have from one up to almost 11,000 workers.

TABLE 8.1 Descriptive Statistics of the Selected Variables (n = 2,349)

Variable	Average	Std. Dev.	Min.	Max.
Total staff (#)	48.43	300.02	1	10,981
Net length (km)	176.79	1,207.54	1	33,180
Collected water (thousands of m^3)	5,281.43	34,274.87	1	1,197,013
Expenses on goods and services ($)	9,830.03	97,872.36	1	3,694,379
Total domestic connections (#)	9,595.85	66,452.10	1	2,379,440

Source: Own elaboration with data from the 2014 Economic Censuses.

The model is then defined by four input variables (total staff, length of networks, volume of water collected, expenses on goods and services), and one output (total domestic connections). This is an input-oriented model, which is a reasonable approach due to the usual shortage of resources in Mexico OCTSAs and their attempts to optimize the use of factors. For the second specification, in the results section, we test and reject the assumption of CRS technology, and the model is specified for VRS. The size of the database (2,349 OCTSA) and the estimation of efficiency with robust non-parametric methods require intensive calculations that, fortunately, can be carried out using different statistical programs[4]. In this work, the five estimators previously defined were run using programs implemented in Stata 15 software.

8.4 RESULTS OF THE NONPARAMETRIC ESTIMATORS

The first step was to detect the presence of outliers[5] in the OCTSAs database. These atypical values occur due to unintentional errors or changes in the data recordings. While the total number of personnel and the length of the networks do not appear to be so dispersed, the variables volume of collected water, expenditure on goods and services, and domestic connections report extremely high values (Figure 1(A) in the Annexure). These influential values are important in the efficiency assessment, especially in the so-called deterministic estimators DEA and FDH.

To find the returns to scale in the model, a test was applied using simulation (bootstrap). Based on Simar & Wilson (2002); and Badunenko & Mozharovskyi (2016), we test the hypothesis: Ho: The technology is CRS. If H_0 is rejected, apply a second test: H_0 = non-increasing returns to scale (NIRS), vs H_1 = variable returns to scale (VRS). Due to the size of the database and the intensity of calculations, the routine of each test run for several hours[6]. The results are reported in Table 8.2.

TABLE 8.2 Results of the Scale Performance Test (200 Repetitions)

	Test	Static Value	p-Value
Test 1	• $\mathbf{H_0}$: Technology (T) is globally CRS.	$S_1 = 0.4030$	p = 0.0000
	• $\mathbf{H_1}$: T is VRS.		
Test 2	• $\mathbf{H_0}$: T is globally NIRS.	$S_2 = 0.4490$	p = 0.0000
	• $\mathbf{H_1}$: T is VRS.		

Source: Own elaboration with data from the 2014 Economic Censuses.

[4] A review of statistical program for nonparametric analysis is found in Daraio et al. (2019).

[5] There is no standard definition, but we take it as extreme values in the sample, with a low probability of occurrence (Fox et al., 2004).

[6] For details of the programs used, *see* Badunenko & Mozharovskyi (2016); and Tauchmann (2012).

In Test 1, the statistical value S_1 represents the ratio between the average measures of technical efficiency assuming VRS and CRS technologies. If H_0 is not rejected, it indicates that the average distance between the VRS and CRS boundaries is small. On the other hand, H_1 suggests the opposite, indicating a large distance. However, if H_0 is rejected, Test 2 is conducted. H_0 assumes that the average distance between VRS and NIRS boundaries is small. H_0 will be rejected if S_2 is significantly greater than 1 (Badunenko & Mozharovskyi, 2016). Based on the results, both S_1 and S_2 are significantly less than 1, indicating that OCTSAs operate under VRS technology. Another finding is that 68.7% of OCTSAs in Mexico are inefficient in scale, while only 31.3% (736) operate efficiently in scale. With these test results, we assume VRS for the estimation of the nonparametric models.

First, the DEA and FDH efficiency estimators are reported. Table 8.3 and Figure 8.2 show that the average efficiency of the FDH estimator is higher than the DEA estimator. This is due to the relaxation of the strict convexity assumption, which allows the efficiency frontier to be based solely on the observed values. Additionally, it was found that the proportion of efficient units on the production frontier (estimators = 1) increases from 275 (11.7%) to 465 (19.8%). The asymmetry and kurtosis values, as well as visual inspection on the graphs, indicate right skewness distribution, although such asymmetry in FDH is reduced. Thus, it looks reasonable that DEA and FDH estimators are biased due to outliers or errors in the data.

TABLE 8.3 Descriptive Statistics of DEA and FDH Estimators (N = 2,349)

Estimator	Mean	Std. Dev.	Skewness	Kurtosis	OCTSA Efficient Units
DEA	0.34924	0.28903	1.25158	3.41406	11.7%
FDH	0.49987	0.30558	0.55080	1.94927	19.8%

Source: Own elaboration with data from the 2014 Economic Censuses.

The first alternative to DEA and FDH estimators is the order-M estimator. Here, while there is no M parameter defined a priori, different values are tested in a range of 100 to 500. That is, each OCTSA was compared with random samples of 100 to 500 units, with 200 repetitions, to obtain its efficiency estimator (θ). Unlike DEA and FDH, the M-order may produce efficiency estimators greater than 1, known as super-efficient values. However, since there are no pre-defined criteria to define the concept of "super efficiency," we chose a value of 1.5 so that an OCTSA is considered an outlier. At 1.5, an OCTSA uses at least 50% fewer inputs than those used in a comparative subsample M to produce a given level of output. In Figure 8.3, with m = 50, one-third of the OCTSAs are super-efficient (score > 1.5); whereas if m = 500, the proportion is reduced to 5%. As expected, the proportion of super-efficient OCTSAs tends to be equal to the proportion of efficient OCTSAs obtained with FDH when $m \to \infty$, which is a consistent result with theory. Indeed, a criterion to identify true outliers is to verify their stability as super-efficient, even with high values of m. In this case, at least 5% of OCTSAs are super-efficient with values of m = 500 and $\theta > 1.5$.

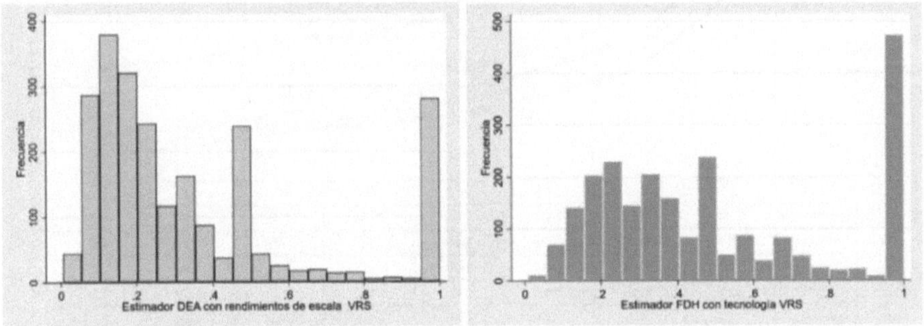

FIGURE 8.2 Histograms of the DEA and FDH efficiency estimators.

Source: Own elaboration with data from the 2014 Economic Censuses.

The alternative to the order-M estimator is the α-order estimator, which is different. It utilizes quartiles (1-α) to assess the efficiency of OCTSAs, by testing various alpha values (ranging from 0 to 1) to examine the impact of outliers on the super efficiency ratio. Figure 8.4 illustrates the percentages of super-efficient OCTSAs (θ> 1.5) at different α values. As α exceeds 0.8, the proportion of super-efficient OCTSAs rapidly declines. It is worth noting that when α = 1, the estimator is equivalent to the FDH estimator. The estimated efficiency using the order-α estimator is extremely high, with the lowest being 1.2 when α = 1. This suggests that a significant proportion of OCTSAs lie outside the production frontier.

Until now, it has been observed how estimators with partial boundaries can assist in identifying outliers. By defining efficiency criteria with a limit > 1.5 for both m-order=400 and α-order=0.8, a total of 145 outliers were identified. One possible approach is to eliminate these outliers and then estimate an unbiased FDH. However, this option would result in a loss of

6% of the data. Another limitation is that the Economic Censuses adhere to principles of confidentiality and data protection, so individual identification or knowledge of each OCTSA to be eliminated is not available. As an alternative, we have chosen to apply the bias-corrected estimator (BC).

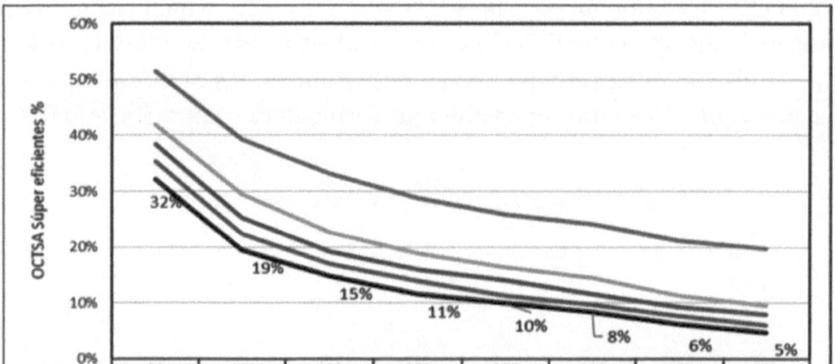

	50	100	150	200	250	300	400	500
>1.5	32%	19%	15%	11%	10%	8%	6%	5%
>1.4	35%	22%	17%	14%	11%	10%	8%	6%
>1.3	38%	25%	19%	16%	14%	12%	9%	8%
>1.2	42%	29%	23%	19%	16%	14%	11%	10%
>1	52%	39%	33%	29%	26%	24%	21%	20%

>1.5 >1.4 >1.3 >1.2 >1

FIGURE 8.3 Proportion of super-efficient OCTSAs with different M-order values and super efficiency levels N = 2,349.

Source: Own elaboration with data from the 2014 Economic Censuses.

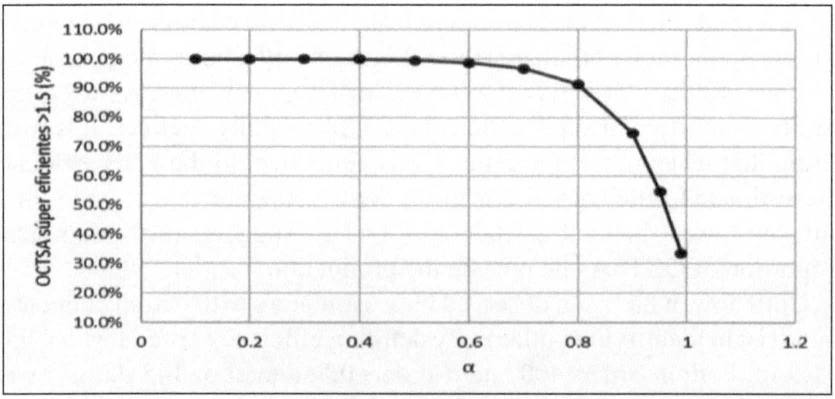

FIGURE 8.4 Super-efficient OCTSA (θ> 1.5) with different values of α (n = 2,349).

Source: Own elaboration with data from the 2014 Economic Censuses.

To estimate BC, the first step is to conduct an independence test of the efficiency estimates, using 999 repetitions and a 95% confidence level. The null hypothesis (H_0) is that the estimated efficiency measures are independent. This test also determines the type of bootstrap to be used (Badunenko & Mozharovskyi, 2006)[7]. The result of this independence test (0.3794) yielded a p-value of 0.0000, indicating rejection of independence and suggesting the use of a homogeneous bootstrap. Therefore, for the BC estimation, we employed a smoothed homogeneous bootstrap with 999 repetitions, assuming a VRS in an input-oriented model.

The second step involved checking the bias correction for outliers. An important finding was that the average efficiency estimated by the BC method for the 145 detected outliers (by m-order and α-order) was 0.69. This clearly indicates the correction achieved by the BC method (Table 8.4).

TABLE 8.4 BC Efficiency Estimates for the OCTSA Group with Outliers

Variable	Obs.	Mean	Std. Dev.	Min.	Max.
Efficiency BC	145	0.69	0.23	0.21	1

Source: Self-made.

Finally, regarding the BC efficiency estimates, the first notable result (Table 8.5) is that the global proportion of efficient OCTSAs (7.7%) decreases significantly when compared to the DEA (11.7%) and FDH (19.8%). The average efficiency for the 182 OCTSAs classified as efficient is 0.31. A second notable result is that the BC method achieves bias correction, reducing the number of super-efficient OCTSAs identified by m-order and α-order criteria from 145 to only 23.

If we organize the BC estimators by the population of OCTSAs served or city size, interesting results emerge. From Table 8.5, even though the BC estimator falls into the largest population group, e.g., more than 1 million inhabitants, it records the best average (0.50), and only the smallest city size reports 100% efficient units.

Figure 8.5 illustrates the last estimated efficiency by city size. A relevant point is that the distributions of the smallest (less than 2,500 inhabitants)

[7] The command used to estimate BC indicators (teradialbc) provides statistical inference for three types of bootstrapping: (i) homogeneous smoothing; (ii) smoothed heterogeneously; and (iii) subsampling (heterogeneous). In this case, we apply to Type 1, following the results of the independence test.

and the largest (more than 1 million) are the ones that differ the most from the other three groups.

TABLE 8.5 BC Estimators by OCTSA Population Served and City Size Group

	<= 2,500 pop	2,500–20,000 pop	20,001–50,000	50,001–1 million	>1 million	Total
Number	798	1,132	210	199	10	2,349
BC Average	0.43	0.27	0.17	0.23	0.5	0.31
Efficient (#)	125	57	0	0	0	182
Efficient (%)	15.6	5.03	0	0	0	7.7%

Source: Own elaboration with data from the 2014 Economic Censuses.

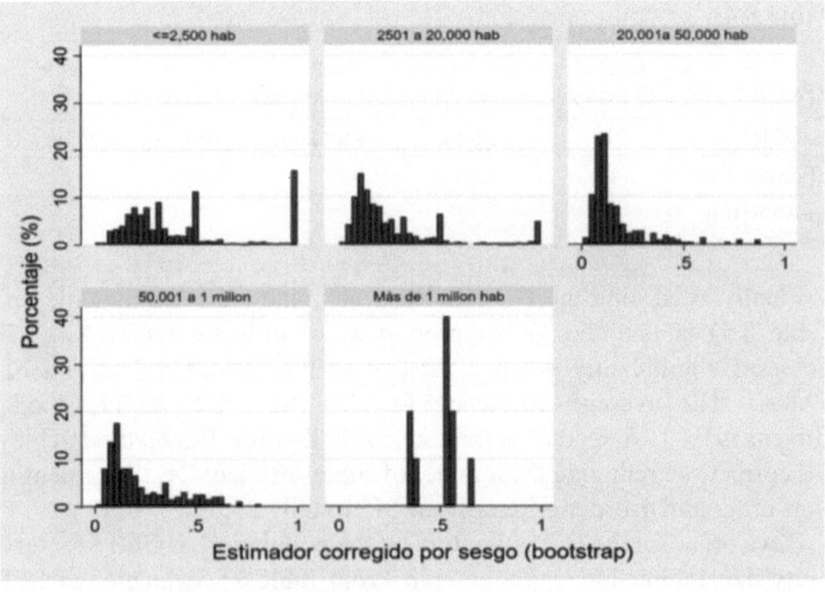

FIGURE 8.5 Histograms of OCTSA BC estimators by population OCTSA served: City size.
Source: Own elaboration with data from the 2014 Economic Censuses.

To conclude, the results obtained from the BC estimator may allow for further analysis, such as examining the impact of external variables on technical efficiency (e.g., water sources, subsidies, water availability, private participation, among others). These topics are suggested for future research.

8.5 CONCLUSIONS

This work demonstrates the potential of the microdata database gathered from the questionnaire applied to OCTSA, as part of the Economic Censuses in Mexico, for the analysis of technical efficiency in water services. By applying five non-parametric efficiency estimators, we conclude that the BC estimator is the one that best manages the bias generated by the presence of outliers and errors in the data. As expected, deterministic estimators (DEA and FDH) have an overestimation bias of efficient units due to the existence of outliers. One test to identify these outliers in the OCTSA database can be based on partial borders methods: m-order and α-order.

As a practical outcome, this research provides an opportunity to identify the best practices that influence efficiency and contribute to improving data collection. This is particularly important in order to prevent inaccurate data reports. Another noteworthy observation, relevant for public policy, is derived from the scale returns test. It is observed that at least one-third of OCTSAs in Mexico currently operate at an inefficient scale. Once these OCTSAs have been identified, operational measures can be taken to adjust their size by managing aspects such as those included in this analysis, such as expenses on goods and services and personnel size.

The results of the bias-corrected estimators (BC) confirm the common perception that the largest OCTSAs (serving a population greater than 1 million) are the most efficient group (7 out of 10 have an estimator greater than 0.50). However, none of them, when compared to the entire OCTSA population, are considered 100% efficient. Another result is observed for the rural OCTSA group (with less than 2,500 inhabitants), where a higher proportion (15%) is efficient. In clear contrast, the majority of OCTSAs in the four groups with less than 1 million inhabitants have low technical efficiencies (less than 0.5). This is an important finding as it suggests that the best performance in practices related to the variables included in this model (expenses on goods and services, total personnel, volume of water collected, and length of networks) is occurring in rural OCTSAs.

Technical efficiency should be one of the criteria for the allocation of public resources. In Mexico, the policy for improving the coverage of basic water services is implemented through the Drinking Water, Drainage, and Treatment Program (PROAGUA). This program is crucial for establishing priorities and defining the Declaration of Priority Attention Areas, which helps introduce measures to make progress in municipalities and localities with high poverty rates and significant social challenges. Other prioritization criteria of PROAGUA include the number of inhabitants, actions aligned with human rights recommendations, actions that contribute to energy saving, among others. However, we believe that incorporating efficiency measures, as estimated in this study, can enhance the certainty in the use of public resources and also facilitate better water management. What if we apply an economic incentive? This means providing rewards or incentives to OCTSAs that demonstrate greater efficiency in resource utilization, including financial management (e.g., expenditure on goods and inputs, including energy), water usage efficiency (e.g., efficient use of captured volumes), and employment. Fortunately, the list of economic instruments or policy options is extensive, as are the experiences and lessons learned from implementing these measures worldwide.

KEYWORDS

- **biased corrected estimator**
- **data envelopment analysis**
- **efficiency frontiers**
- **free disposal hull**
- **robust non-parametric techniques**
- **water utilities**

REFERENCES

Aguilar-Benítez, I., (2011). Viabilidad financiera de los servicios del agua: Una comparación de tres pares de ciudades en la frontera México-Estados Unidos. En: Aguilar-Benítez, I., (coord.), *Los Servicios del Agua en el Norte de México: Gestión, Manejo Financiero y Aspectos Socio-Ambientales*. Tijuana, Colef/Colson.

Ananda, J., (2014). Evaluating the performance of urban water utilities: A robust nonparametric approach. *Journal of Water Resources Planning and Management, 140*(9). doi: 10.1061/(asce)wr.1943-5452.0000387.

Anwandter, L., & Ozuna, T. Jr., (2002). Can public sector reforms improve the efficiency of public water utilities? *Environment and Development Economics, 7*(4). doi: 10.1017/s1355770x02000414.

Aragon, Y., Daouia, A., & Thomas-Agnan, C., (2005). Nonparametric frontier estimation: A conditional quantile-based approach. *Econometric Theory, 21*(2), 358–389. Retrieved from www.jstor.org/stable/3533469 (accessed on 24 February 2024).

Badunenko, O., & Pavlo, M., (2016). Nonparametric frontier analysis using Stata. *The Stata Journal, 16*(3), 550–589. doi: 10.1177/1536867X1601600302.

Banker, R. D., Charnes, A., & Cooper, W. W., (1984). Some models for estimating technical and scale inefficiencies in data envelopment analysis. *Management Science, 30*(9), 1078–1092.

Bowlin, W., (1998). Measuring performance: An introduction to data envelopment analysis (DEA), *The Journal of Cost Analysis, 15*(2), 3–27. doi: 10.1080/08823871.1998.10462318.

Cabrera, E., Estruch-Juan, E., & Molinos-Senante, M., (2018). Adequacy of DEA as a regulatory tool in the water sector. The impact of data uncertainty. *Environmental Science & Policy, 85*, 155–162. doi: 10.1016/j.envsci.2018.03.028.

Cazals, C., Florens, & Simar, (2002). Nonparametric frontier estimation: a robust approach. *Journal of Econometrics, 106*(1), 1–25.

Charnes, Cooper, W. W., & Rhodes, E., (1978). Measuring the efficiency of decision-making units. *European Journal of Operational Research*, 429–444.

Cooper, W., Seiford, L. M., & Zhu, (2011). *Handbook on Data Envelopment Analysis; International Series in Operations Research & Management Science* (2nd edn., Vol. 164).

Daouia, A., & Simar, L., (2007). Nonparametric efficiency analysis: A multivariate conditional quantile approach. *Journal of Econometrics, 140*(2), 375–400. doi: 10.1016/j.jeconom.2006.07.002.

Daraio, C., & Simar, L., (2007). *Advanced Robust and Nonparametric Methods in Efficiency Analysis: Methodology and Applications*. Springer. 10.1007/978-0-387-35231-2.

Daraio, et al., (2019). Productivity and efficiency analysis software: An exploratory bibliographical survey of the options. *Journal of Economic Surveys, 33*(1), 85–100.

De Witte, K., & Marques, R. C., (2010). Influential observations in frontier models, a robust non-oriented approach to the water sector. *Annals of Operations Research, 181*(1), 377–392. doi: 10.1007/s10479-010-0754-6.

Deprins, D., & Tulkens, H., (1984). Measuring labor efficiency in post offices. In: Marchand, M., & Tulkens, H., (eds.), *The Performance of Public Enterprises: Concepts and Measurement North-Holland* (pp. 243–267).

Fox, K., Hill, R., & Diewert, W., (2004). Identifying outliers in multi-output models. *Journal of Productivity Analysis, 22*(1, 2), 73–94.

Griffin, R. C., (2016). *Water Resource Economics: The Analysis of Scarcity, Policies, and Projects.* Cambridge: MIT Press.

Güngör-Demirci, G., Lee, J., & Keck, J., (2018). Measuring water utility performance using nonparametric linear programming. *Civil Engineering and Environmental Systems, 34*(3, 4), 206–220. doi: 10.1080/10286608.2018.1425403.

Kneip, A., Simar, L., & Wilson, P., (2008). Asymptotics and consistent bootstraps for DEA estimators in nonparametric frontier models. *Econometric Theory, 24*(6), 1663–1697.

Marques, R. C., & De Witte, K., (2011). Is big better? On scale and scope economies in the Portuguese water sector. *Economic Modelling, 28*(3), 1009–1016. doi: 10.1016/j.econmod.2010.11.014.

Molinos-Senante, M., Donoso, G., & Sala-Garrido, R., (2016). Assessing the efficiency of Chilean water and sewerage companies accounting for uncertainty. *Environmental Science & Policy, 61,* 116–123. doi: 10.1016/j.envsci.2016.04.003.

Picazo-Tadeo, A. J., González-Gómez, F., & Sáez-Fernández, F. J., (2009). Accounting for operating environments in measuring water utilities' managerial efficiency. *The Service Industries Journal, 29*(6), 761–773. doi: 10.1080/02642060802190300.

Sampaio De, S., & Stosic, B., (2005). Technical efficiency of the Brazilian municipalities: Correcting nonparametric frontier measurements for outliers. *JPA, 24*(2), 157–181.

Simar, L., & Wilson, P. W., (2002). Non-parametric tests of returns to scale. *European Journal of Operational Research, 139*(1), 115–132.

Tauchmann, H., (2012). Partial frontier efficiency analysis. *The Stata Journal, 12*(3), 461–478. doi: 10.1177/1536867X1201200309.

Worthington, A. C., (2013). A review of frontier approaches to efficiency and productivity measurement in urban water utilities. *Urban Water Journal, 11*(1), 55–73. doi: 10.1080/1573062x.2013.765488.

ANNEXTURE

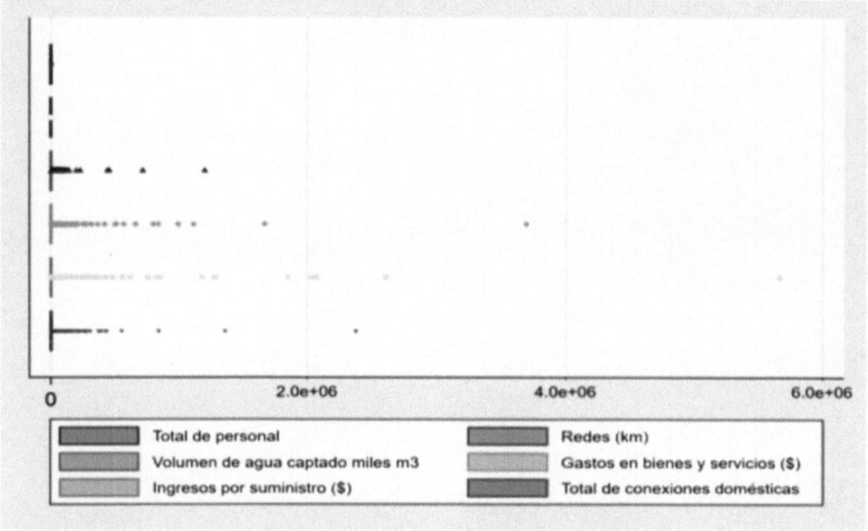

FIGURE 1(A) Main variables of the OCTSA model (n = 2,349).

Source: Own elaboration with data from the 2014 Economic Censuses.

CHAPTER 9

Water Quality Index: Measure of Quality of Drinking Water for Sustainable Development

PARVATHI JAYASANKAR,[1] K. V. SATHEESH KUMAR,[2] V. VEDAVALLI,[3] AGAMPODI SANDALI INUPAMA DE ZOYSA,[3] and U. MEGHANA[3]

[1]*Department of Chemistry and Biochemistry, JAIN (Deemed to be University), Bangalore, Karnataka, India*

[2]*Postgraduate and Research Department of Chemistry, Government College, Kasaragod, Kerala, India*

[3]*JAIN (Deemed to be University), Bangalore, Karnataka, India*

ABSTRACT

Water is an invisible natural resource that is facing global deterioration. The measurement of the quality of this vital natural resource is important. The water quality is deteriorating due to various anthropogenic activities such as urbanization, unconventional agricultural activities, and other socioeconomic developments. It is scientifically recommended that parameters for water quality assessment be developed. Similar to the UV index, the water quality index (WQI) is a summarized data of various water quality parameters documented for reporting to the public. These parameters include pH, alkalinity, acidity, TDS, chlorine content, total hardness, etc. The various physicochemical parameters of the water samples were analyzed and assigned an average weightage (Wi) for each parameter with respect to its relative importance in the overall quality of

Sustainable Environment and Health: Practical Strategies. Leonardo Sepúlveda Torre, Ann Rose Abraham, Neha Kanwar Rawat, & A. K. Haghi (Eds.)

portable drinking water. Furthermore, the relative Wi is calculated. At the end, a quality rating scale is assigned for each parameter, taking into account national/international standards. Then, the subindex of the ith parameter is calculated, which, upon integration, gives the value of the WQI.

The computed WQI values are classified into five types: excellent, good, medium, bad, and very bad water quality. These classifications facilitate the need to protect the water from various contaminants and also provide awareness about the level of treatment required to achieve the desired drinking quality for the water sample.

9.1 INTRODUCTION

9.1.1 IMPORTANCE OF QUALITY OF WATER

Water is the elixir of human life, and water quality is a significant criterion in meeting the demands of a population with the available water resources in the area. Adequate water with the required quality is a fundamental prerequisite of a civilized society and is considered an important aspect of environmental management and sustainable development goals (SDGs). Sustainable water resource management is essential for the socioeconomic growth of a country. India has scarce water resources to sustain its vast population. The majority of farming in India relies on untreated water, which contaminates food products and leads to serious health issues for farmers. The motivation for farmers to use wastewater for their crops is a result of various national challenges, including rapid population growth and urbanization, increasing demand for food, and severely depleted reserves of clean water (Chaturvedi et al., 2020). The quality of life for humans and the ecosystem of an area or country mainly depends on the availability of water of the right quality and how the population uses and maintains that quality on a continuous basis.

Both surface water and groundwater used in municipal water must be continuously monitored for quality for industrial, agricultural, domestic uses, and for ecosystem health and hygiene (Figure 9.1). When determining the quality of water, we need to monitor the physical, chemical, and biological aspects of water (Figure 9.2). All the above parameters may not be directly impacted by pollution or anthropogenic activities. They may also be impacted by erosion, climatic conditions, and the regeneration of water by rain or snow.

FIGURE 9.1 Monitoring of quality for industrial, agricultural, domestic uses, and for ecosystem health and hygiene.

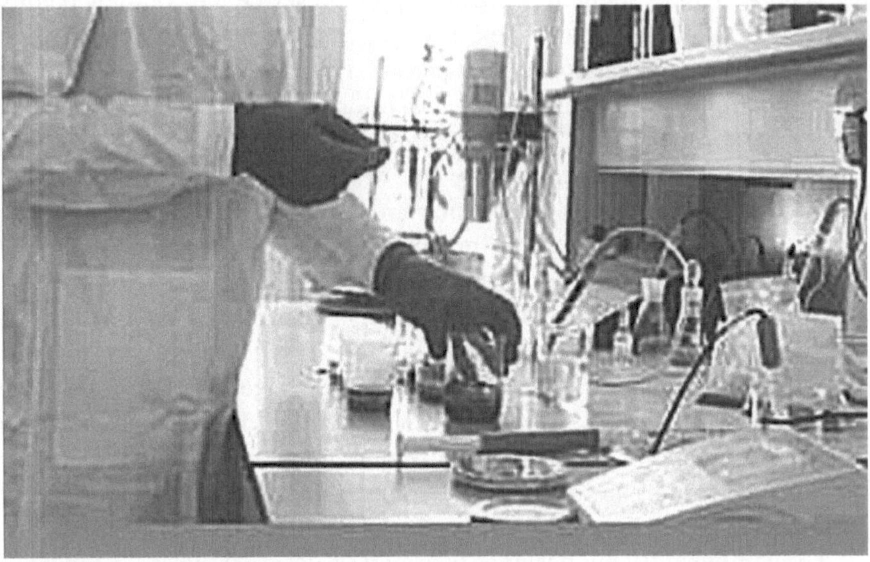

FIGURE 9.2 Monitoring of the physical, chemical, and biological aspects of water.

9.2 INTRODUCTION TO WATER QUALITY INDEX (WQI) AND ITS ADVANTAGES FOR WATER QUALITY MONITORING

The water quality index (WQI) is a single value that provides an understanding of the overall quality of water at a specific site and time period, based on various water quality parameters. The main purpose of this index is to easily communicate the water quality to the public. It serves as an indicator of water quality, presenting complex data in a comprehensible manner (http://bcn.boulder.co.us/basin/watershed/wqi_info.html). The process of monitoring and managing water quality begins with determining the average values of physical, chemical, and biological parameters. These parameters are scientifically tested in laboratory environments after sample collection. Instead of analyzing the results individually, scientists and researchers have developed a composite index to compare and categorize the quality into different levels.

The WQI helps us provide a summarized report on the nature of water quality to the public in a consistent manner. It is formulated by averaging the index values of selected water quality parameters over a period of time. The objective of the index is to transform complex water quality measurements and comparisons into information that is understandable and usable by the general public. While it is challenging to describe the complex quality scenario of water using an index, it provides a relative measure of water quality for a specific area and time period. The objective of the WQI is to convert complex water quality data into information that is understandable and usable for the general public (Inayathulla & Jai, 2013).

While arriving at an index, the first task is to determine the number of physiochemical parameters which are important for determining the quality of the water intended for potential use. The readings of the quality parameters are given weighted importance and grading vis-a-vis the available standards for each of the parameters. Although there are different methods used for the calculation of the index, which is briefly touched upon in the subsequent section, the index gives a weighted average score of the composite quality of water. This index provides aggregate data on various water quality parameters at different times and places and converts the whole information into a single value, giving information about that particular space at a particular time (Abdul, Bahram, & Abass, 2010).

The computed WQI is designed to provide a clear ranking of water quality on a scale from 0 to 100. Higher values of the WQI indicate better water quality, while lower values indicate poor water quality (Talent & Kumarasamy, 2020). The integrated water quality index (IWQI) is used because the concentration of any parameter below the desirable limit or above the permissible limit contributes to an increase in the overall index value (Mukathe, Vagh, Panaskar, Jacobs, & Sawanth, 2019). Various methods have been historically adopted worldwide to determine WQI. These methods are preferred over individual testing methods for the following reasons:

- Water sources are not always consistent in terms of quality, making conventional testing methods impractical or inconclusive.
- Pollution levels in a water body can vary across different sections due to different uses of water. In such cases, a relative index of water quality is useful in determining the overall quality of water across the body.
- The index provides a quick overview for the public and policy holders to assess the quality of different areas of the water body or changes in water quality over time. However, the WQI should be used with caution and diligence, depending on the intended purpose. The selection of parameters (ranging from 5 to 18) is generally used to calculate a composite WQI. Additionally, the weighting of different parameters must be carefully chosen based on the desired objective. Certain confined water bodies, such as ponds, may not be suitable for effectively using WQI to determine water quality.

9.3 HISTORIC USE OF WQI INDEXES AND METHODOLOGY

Historically, WQI has been used in various forms and shapes since 1965. Horton (Uddin, Moniruzzaman, & Khan, 2017) initially used WQI calculations in the US, which were later modified and strengthened by Brown and collaborators in NFS-WQI (Lumb, Sharma, & Bibeault, 2011). The Environmental Department of Oregon and the Canadian Council of Ministers developed different versions of WQI in 1970 and 1990, respectively (Calmuc et al., 2018). The weighted arithmetic WQI method is a suitable method for determining the WQI as it provides a composite WQI that assigns weighted importance to the parameters mentioned. Therefore, numerous methods for calculating WQI exist in different parts of the world, depending on their specific needs in the area of interest. There is no universally recommended method by experts for worldwide usage, as no method can provide 100% objectivity and accuracy (Calmuc et al., 2018).

Given below are calculations and formulae for WQI calculations with three representative methods:

1. **National Sanitation Foundation–Water Quality Index (NFS WQI):** Nine quality parameters (TDS (total dissolved solids), pH, turbidity, phosphates, nitrates, CBO_5 (biochemical oxygen demand (BOD)), coliforms, DO (dissolved oxygen), and temperature) are measured and assigned a relative weight based on their contribution to calculate the Index using the formula (Mihăiescu, Mihăiescu, Vârban, Vârban, & Mihăiescu, 2013):

$$\text{NFS WQI} = \sum \text{i=1 to n Wi Qi}$$

where; 'Wi' is the weightage; and 'Qi' is the subindex of the quality parameter 'I' obtained from conversion curve.

Once NFS WQI is calculated, ranges are determined to assess the quality as Good, Bad, Excellent, etc.

2. **Oregon Water Quality Index (OWQI):** A derivation of NFS WQI is used where a different method of calculation is employed to determine the index. The key difference is that weightages are not assigned, and an arithmetic average is used as the basis for this calculation. The eight physical, chemical, and biological parameters used in this index calculation are temperature, DO, BOD, pH, ammonia + nitrate nitrogen, total phosphorus, TDS, and coliforms (Cude, 2001).

$$OWQI = \sqrt{\frac{n}{\sum_{i=1}^{n} \frac{1}{SIi^2}}}$$

where 'n' is the number of parameters (total 8 in number); and 'Si' is the sub-index of the 'Ith' parameter. The calculation of the sub-index is explained in the chapter (Cude, 2001). The obtained values of the indexes are then graded on a quality scale ranging from Excellent to Very poor quality of water.

3. **Canadian Council of Ministers of the Environment Index (CCME–WQI):** Used primarily for assessing the quality of water in a stream, the following parameters are tested: temperature,

conductivity, color, turbidity, DO, pH, alkalinity, Ca, Na, Mg, K, SO_4, Cl-, F-, Dissolved Organic Carbon, P, Nitrates, Nitrites, N, SiO_2, Al, As, Ba, Be, Cd, Co, Cr, Cu, Fe, Hg, Li, Mn, Mo, Ni, Pb, Se, Sr, V, Zn (Akovljevic, 2012). Using the experimental results, three main factors, F1 (% of values that do not meet the objectives at least once during the period – *How many*), F2 (percentage of individual tests that did not meet the goals at all – *How often*), F3 (the factor that represents the amount by which the test values did not respect their guideline values – *How much*), are calculated. After determining F1, F2, and F3 through their individual formulas, CCME–WQI is calculated using the formula.

$$F1 = (\text{Number of failed variables/Total number of variables}) \times 100$$

$$F2 = (\text{Number of failed tests/Total number of tests}) \times 100$$

Excursion value – Analytical value exceeding the standard limit

$$\text{Excursion} = (\text{Failed test value/Objective}) - 1$$

Normalized sum excursion (nse) = (Σi excursion i/Total number of tests)

$$F3 = \text{nse}/(0.01\text{nse} + 0.01)$$

$$CWQI = 100 - \sqrt{}(F1^2 + F2^2 + F3^2)/1.732$$

The CWQI, thus calculated, is then categorized into different quality bands ranging from excellent to poor.

4. **Weighted Arithmetic Water Quality Index Method (WAQI):** This method involves conducting experiments and recording the values of various parameters (pH, BOD, COD, DO, $P\text{-}PO_4^{3-}$, N-total, $N\text{-}NO_3^-$, $N\text{-}NO_2^-$, $N\text{-}NH_4^+$, SO_4^{2-}, Cl^-, Cr-total, Pb_2^+, Cd_2^+, Ni_2^+, Fe-total, Mn-total, Zn_2^+, As_2^+). These values are then used to calculate the WAQI using weighted calculations for each parameter.

$$W_i = \left(wi / \sum_1^n wi\right)$$

where; *wi* is the weight of individual parameter and W_i is the relative weight of the parameter.

The indexes thus calculated are then graded based on ranges, resulting in a grading of water quality from Excellent to Unsuitable for drinking.

As mentioned earlier, there is no consensus among experts on the correct final method of calculating WQI. Each method has its own advantages and disadvantages and must be carefully selected depending on the following factors:

- Number of parameters used for testing;
- Values of parameters experimented and calculated;
- Type of waterbody or sources;
- End objective – Whether the quality assessment is for general water quality indexing, assessing the extent of pollution, or assessing the potability of drinking water.

Broad pros and cons of each of the above methods are listed in Table 9.1.

9.4 CONCLUSION

Thus, it is imperative to use the right methods for calculating WQI in order to determine the quality of water. However, the tool and indices are powerful tools for assessing the overall quality of water and communicating it to a large population. Additionally, the index helps us to implement appropriate countermeasures and mitigation tactics for water bodies.

9.4.1 LIMITATIONS IN USING WQI TO DETERMINE WATER QUALITY

There are limitations in using WQI to determine water quality. Although there are many advantages to this technique, WQI may not provide an absolute measure of the degree of pollution in a water body at a given time due to several reasons:

- The WQI models are not generic as they are developed based on site-specific guidelines for a particular geographical region (Kachroud, Trolard, Kefi, Jebari, & Bourrié, 2019).
- Despite using the same variables, the classification from each index differs, especially in the evaluation of drinking water. This difference is mainly due to the variation of anthropogenic activities in the chosen geographical area (Uddin, Md. Nash, & Olbert, 2021).
- A single parameter with a poor value can significantly impact the overall assessment of water quality index (WQI).

TABLE 9.1 Pros and Cons of Various Methods for Calculating WQI

	NFS WQI	OWQI	CCME-WQI	WAQi
Pros	• Fast and objective. • Index determine end water use.	• Can determine most impacted parameter. • Sensitive to environmental factors.	• High adaptability for different end objective. • Low level of sensibility for missing data and hence can be used for automated sampling.	• Suitable for surface water and ground water. • Importance of parameters can be completed. • Easy to understand.
Cons	• Does not use complex scale. • Data loss during calculation.	• Biological factors and impact not assessed.	• No weightages for parameters. • Biological parameters cannot be used.	• Real quality of water is not assessed. • Only determine the pollution level impact.

KEYWORDS

- **anthropogenic activities**
- **degree of pollution**
- **quality of water**
- **total dissolved solids**
- **unconventional agriculture activity**
- **urbanization**
- **water quality index**
- **water quality monitoring**

REFERENCES

Abdul, H. M. J. A., Bahram, K. M., & Abass, J. K., (2010). Evaluating raw and treated water quality of Tigris River within Baghdad by index analysis. *J. Water Resource and Protection, 2*, 629–635.

Akovljevic, D., (2012). Serbian and Canadian water quality index of Danube River in Serbia in 2010. *J. Geogr. Inst. Cvijic., 62*, 1–18.

Calmuc, A. V., Itisecu, C., et al., (2018). *Various Methods for Calculating the Water Quality Index*.

Chaturvedi, V., Kangkanika, N., Sujata, B., Arunabha, G., Sumit, K. G., & Ishita, J., (2020). *Reallocating Water for India's Growth: Sectoral Withdrawals, Water-Efficient Agriculture, and Institutional Mechanisms.* New Delhi: Council on Energy, Environment and Water.

Cude, C. G., (2001). Oregon water quality index a tool for evaluating water quality management effectiveness. *Journal of the American Water Resources Association, 37*(1), 125–137.

http://bcn.boulder.co.us/basin/watershed/wqi_info.html (accessed on 24 February 2024).

Inayathulla, M., & Jai, M. P., (2013). Water quality index assessment of ground water in Jakkur Sub Watershed of Bangalore, Karnataka, INDIA. *International Journal of Civil, Structural, Environmental and Infrastructure Engineering Research and Development (IJCSEIERD)* (Vol. 3, No.1). ISSN 2249-6866.

Kachroud, M., Trolard, F., Kefi, M., Jebari, S., & Bourrié, G., (2019). Water quality indices: Challenges and application limits in the literature. *Water, 11*(2), 361. https://doi.org/10.3390/w11020361.

Lumb, A., Sharma, T. C., & Bibeault, J. F. A., (2011). Review of genesis and evolution of water quality index (WQI) and some future directions. *Water Qual. Expo. Health., 3*, 11–24.

Mihăiescu, T., Mihăiescu, R., Vârban, D., Vârban, R., & Mihăiescu, M., (2013). Analele Universităţii din Oradea. *Fascicula Protecţia Mediului., 11*, 649–654.

Mukathe, S., Vagh, V., Panaskar, D., Jacobs, J. A., & Sawanth, A., (2019). Development of new integrated water quality index (IWQI) model to evaluate the drinking suitability of water. *Ecological Indicators, 101*, 348–354.

Talent, D. B., & Kumarasamy, M., (2020). Application of multivariate statistical analysis in the development of a surrogate water quality index (WQI) for South African watersheds. *Water, 12*, 1584. doi: 10.3390/w12061584. www.mdpi.com/journal/water (accessed on 24 February 2024).

Uddin, M. G., Moniruzzaman, M., & Khan, M., (2017). Evaluation of groundwater quality using CCME water quality index in the rooppur nuclear power plant area, Ishwardi, Pabna, Bangladesh. *American Journal of Environmental Protection, 5*(2), 33–43.

Uddin, Md. G., Nash, S., & Olbert, A., (2021). A review of water quality index models and their use for assessing surface water quality. *Ecological Indicators, 122*, 1470–1486. 10.1016/j.ecolind.2020.107218. doi: 10.1016/j.ecolind.2020.107218.

CHAPTER 10

Era of Managing and Reviving Water Footprint

REKHA MIRLE, AGNESS CHACHA, N. VEDASHREE, PRIYANKA GHOW, and M. KEERTHANA

School of Sciences, Jain University, Bangalore, Karnataka, India

ABSTRACT

In today's developing world, nothing seems impossible to achieve except eliminating human ignorance with respect to sustainability. As a result, we are water. In faced with numerous obstacles for accomplishing simple tasks like saving this chapter we will be focusing on the ways that could be implemented in getting an extra mile in preserving water and water resources. Reuse and recycle of water help to minimize the utilization of water resources and reinforce the exploitation of other substitute resources. With sustainable management of water, we will be able to diminish the risk of global disease burden, improve health and increase overall productivity of society. Moreover, this review provides a brief insight on various risk factors involved in conserving our water resources for upcoming generations, various ways to overcome them which leads managing of water in a sustainable manner.

10.1 INTRODUCTION

Water is a vital part of sustainable development and a crucial factor affecting socio-economic development in fields such as human health,

Sustainable Environment and Health: Practical Strategies. Leonardo Sepúlveda Torre, Ann Rose Abraham, Neha Kanwar Rawat, & A. K. Haghi (Eds.)

economy, environment, agriculture, and overall population. It is important in reducing the global burden of diseases, improving health, and supporting biodiversity. It also serves as a crucial link between climate change, human society, and the environment (Un.Org/waterforlifedecade/water_and_sustainable_development.schtml). The Earth's surface is covered with water, about 70% of which is freshwater, with only 3%–4% being fresh and 1.2% being drinkable. Water sustainability is the process of utilizing water to fulfill our needs and striking a balance with the needs of future generations. The demand for freshwater is escalating, and scientists believe that water shortage in the future is inevitable (Watconman.org). The water footprint refers to the measure of the amount of freshwater utilized to produce goods and services. This water footprint also provides an estimate of the total amount of water consumed in production activities, individual-level services, business-level services, and globally. Water is a sustainable resource that needs protection to meet current and future needs. Therefore, water sustainability and management applications are necessary, as they encompass all the policies, strategies, and activities made to manage and protect water resources (Watconman.org). The increase in population, agriculture, climate change, and industrialization affect the amount of water used, leading to water scarcity. Better water management has social and economic benefits, helping to protect our environment. Thus, water sustainability should be a compulsory practice worldwide.

10.2 DIFFERENT METHODS OF WATER CONSERVATION

We cannot produce water artificially, but we can rely on the available water resources on Earth. Managing water resources involves controlling water flows and storage. Various methods have been observed to yield positive results in water conservation. We will discuss a few of these methods in subsections.

10.1.1 RAINWATER HARVESTING

The method of obtaining and storing rainwater is called rainwater harvesting. Rainwater is usually collected from rooftops through various methods. Rainwater harvesting is not only the simplest among all the methods but also the oldest and the most common method of conserving water (Gosmartbricks.com). Rainwater harvesting dates back to 1700 BC and has been traditionally practiced in regions of India and Pakistan. It was invented independently in different parts of the world later. It is the easiest, most reliable, and least polluted source (Patil et al., 2021). Rainwater harvesting installations are designed based on the infrastructure of the buildings to serve different purposes in schools, hospitals, household activities, and other civil services (Network, 2017). Rainwater from rooftops is suitable for domestic needs such as drinking, irrigation, and livestock management. The collected water can also be used to recharge the groundwater table and store water for a long duration (Kinkade-Levario, 2007). With the growing number of freshwater-flooded forests, it is possible to harvest rainwater efficiently from water-logged land. Rainwater can also be harvested from solar power panels. The water collected from the solar panels is free of contamination with less salt content produced by easy filtration and addition of disinfectant (Dhanraj, Gopi, & Loganathan, 2019; Hanley, 2017; Pham, 2017). Other innovations are rooftop catchment, rain saucer, and free-standing rainwater collection.

Rainwater harvesting is a boon for developing countries that are suffering from water scarcity and helps improve their economy. It also provides water during dry seasons; it's cost-effective and environmentally friendly. The water obtained has negligible contamination and is saline-free. It also reduces the demand for groundwater, thus sustaining its levels (Figure 10.1).

FIGURE 10.1 Rain water harvesting model.

10.1.2 GROUND WATER RECHARGE

This refers to a hydrological process that involves the movement of surface water into the ground. This water is then referred to as groundwater. Groundwater recharge is made possible due to the presence of aquifers, and its movement occurs in a region below the plant roots. The movement of water from the water table to a region of porous rocks filled with water (saturated zone) is also referred to as groundwater recharge (Freeze, 1994). Groundwater recharge can occur naturally or through human activities. Naturally, the process of recharging groundwater is aided by natural water sources such as surface water (rivers, lakes), rain, and in some cases, snow. The process of recharge can be interfered with by human activities such as agriculture and road constructions, which result in the loss of the fertile soil layer, reducing the penetration of water into the ground. The process helps in the leaching of excess salts from roots into other soil layers, and the presence of trees in an area helps minimize water runoff (Szatko, Tang, Tian, & Rodie, 2012). Even during rainy seasons, waterlogged areas help in the process of groundwater recharge (Major Floods Recharge Aquifers, 2011). Climate change, urbanization, wastelands due to agricultural processes, constructions, and pavement development are some factors that affect groundwater recharge.

10.1.3 SEWAGE WATER TREATMENT

It is the process in which wastewater from households, industries, and business centers is treated to remove contaminants for the purpose of reuse. The treated water is commonly used in the agricultural sector for irrigation or even in construction sites.Sewage treatment of water mainly involves four processes:

i. **Pretreatment:** Removal of large solid particles that can cause blockage.
ii. **Primary Treatment:** Elimination of suspended solids and organic matter.
iii. **Secondary Treatment:** Use of microorganisms to eliminate the organic matter causing pollution.
iv. **Tertiary Treatment:** Nutrient removal and disinfection.

Wastewater treatment, also known as water reclamation, serves the main purpose of reducing environmental pollution and distributing water to areas with water scarcity. It can also be used for drinking purposes and as a solvent for industrial purposes (Figure 10.2).

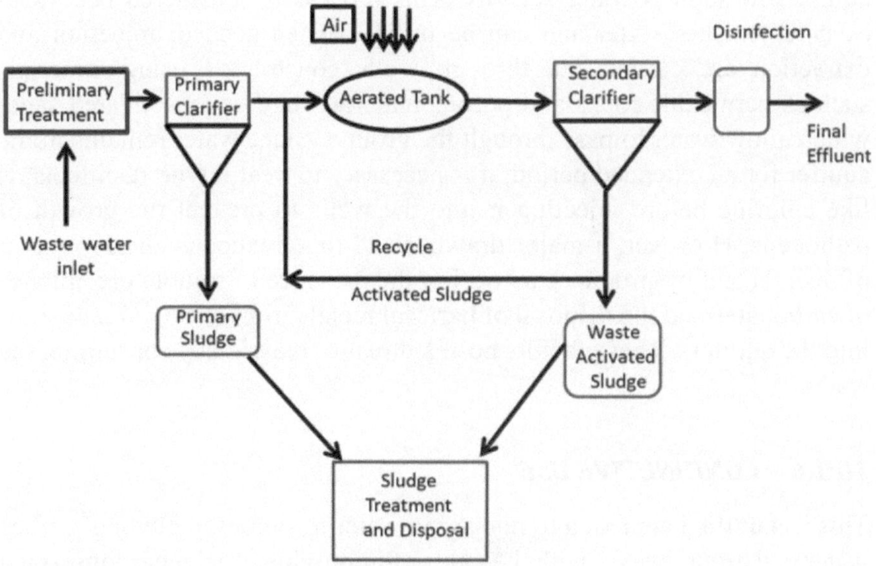

FIGURE 10.2 Flowchart of sewage water treatment process.

Source: Adapted from Azadi et al., 2015

10.1.4 DESALINATION

Desalination is the process of removing dissolved salts from seawater, which serves as a major source of saline water (Dictionary, 2005). Desalinated water can be used for various purposes, including irrigation, industrial activities, and domestic use (Panagopoulos, Haralambous, & Loizidou, 2019). The primary objective of this process is to provide a cost-effective solution to ensure the availability of water sources for human consumption (Fischetti, 2007). Desalinating seawater requires a significant amount of energy and is generally more expensive. However, desalination is a reliable technique that is not at risk of depletion like other sustainable water resources (Ebrahimi, Najafpour, & Kebria, 2018; Buell, Anderson, De Pablo, & Oskenbay, 2020). Desalination technology utilizes a large amount of energy, which can be obtained from fossil fuels through thermal or mechanical evaporation processes (Cohen, 2021).

10.1.5 AQUIFER STORAGE AND RECOVERY

Aquifer storage is an intentional process of storing water in underground aquifers to address water scarcity. This process is considered beneficial as it replenishes water and can be utilized when needed. Injection and extraction are carried out through wells constructed using materials such as permeable rocks and porous materials like unconsolidated sand, which allow water to pass through the ground. Since water remains in the aquifer for an extended period, it is necessary to treat it with disinfectants like chlorine before injecting it into the wells to prevent the growth of pathogens. However, a major drawback of this method is the formation of disinfectant by-products, as well as the presence of soluble precipitates of carbonates and the disposal of harmful metals from chemical industries into the aquifers. These factors pose a threat to individuals consuming the water.

10.1.6 CONJUNCTIVE USE

This is a unified approach to managing water resources, including surface water and groundwater, with the goals of improving efficiency, conserving water for long-term use, and mitigating negative impacts on various water sources. Ideally, it optimizes the use of both surface and groundwater during wet and dry seasons. Conjunctive management programs are implemented in various parts of the world to control declining groundwater levels and water scarcity. These programs can be implemented at local or global scales.

Advantages of conjunctive management:

- It enhances the availability of water to local areas;
- It reduces overflow of groundwater;
- It reduces the chances of water logging;
- It helps to achieve our socio-environmental needs.

10.2 IMPACTS OF SUSTAINABLE MANAGEMENT OF WATER ON SOCIETY

The overall aim of water sustainability is to ensure that the world no longer suffers from water scarcity in years to come. Therefore, different methods are implemented in many countries, especially developing countries. Water plays a crucial role in various sectors of our lives. Managing the use of water sensibly helps to overcome other threats that the world might face in the future, such as food shortage and the sustenance of animal and plant life on Earth. The impacts of water sustainability on society are discussed briefly below:

1. **Agriculture:** It is the cornerstone of any country, and the world depends on it for survival. Agriculture is the largest consumer of water, and a significant amount of water is wasted and used to support agricultural activities. This highlights the importance of utilizing methods that reduce the daily water consumption in agriculture. Techniques such as using drought-resistant crops, capturing and storing rainwater in wells for irrigation, and adjusting crop growth based on soil moisture content during dry seasons are crucial (Shamsuddin et al., 2014).

2. **Drip Irrigation:** This method has been successful in reducing water utilization in agriculture. It minimizes water loss and evaporation by delivering water directly to the crop roots in the required amount. As the largest consumer of water, the agricultural sector greatly benefits from sustainable management methods that offer various techniques for crop modifications to adapt to different situations (Obaideen et al., 2022).

3. **The Industrial Sector:** Industrial waste is one of the major pollutants that induces the risk of environmental pollution. Hence, treating wastewater plays a significant role in minimizing water waste throughout the process, as it can be recycled and reused, thereby reducing water loss. Moreover, sewage treatment helps to recover nutrients that can be potential raw materials for other processes (Weerasekara, 2017; Grizzetti et al., 2016).

4. **Domestic Use:** Water sustainability and management methods, such as rainwater harvesting, have been seen to provide a positive outcome as a water source. Water harvested from rain can be stored and later used for household purposes such as drinking, cooking, gardening, and other chores. Due to its easy accessibility and reliability, it improves water availability for

domestic use even during dry seasons. For areas facing water shortage, aquifer storage has been seen to be of great help in providing individuals with water for domestic use (Clasen, 2015).

5. **Ecosystem:** There are different ecosystems on Earth, and they all depend on one crucial compound called water. In an ecosystem, water is essential for the growth and development of plants and other species, acts as a breeding ground for some organisms, and provides a habitat for aquatic organisms, such as the marine ecosystem (Postel & Richter, 2012). The availability of water leads to biodiversity, which increases the richness of species in a particular region. Due to its distinct characteristics, water is a unique solvent, and it is impossible for the existence of another substance that can substitute water. Thus, water sustainability is important for the ecosystem (Boelee, 2011).

10.3 CONCLUSION

Water is an essential resource that supports life on Earth. Due to the population explosion, the world is facing major challenges regarding water safety, control of water pollution, and providing a healthy environment for the ecosystem. Climate change and water diversions threaten the environment and ecosystem; however, the presence of traditional methods such as rainwater harvesting helps to minimize the effect to a smaller extent (Cosgrove & Cosgrove, 2012). Provision should be made to implement policies to achieve sustainable water management goals. The predictable increased variability, availability, and distribution of freshwater resources demand political commitment in supporting and advancing technology for the collection and analysis of hydrological data. More up-to-date information will enable policymakers to make better decisions regarding water resource management, leading to sustainability (Cosgrove & Cosgrove, 2012).

KEYWORDS

- **conservation**
- **desalination**
- **rainwater harvesting**
- **risk factors**
- **sewage water treatment**
- **society**
- **sustainability**
- **sustainable management**
- **water footprint**

REFERENCES

Azadi, Ali & Fallahzadeh, Reza & Sadeghi, Shahram. (2015). Dairy wastewater treatment plant in removal of organic pollution: a case study in Sanandaj, Iran. Environmental Health Engineering and Management. 2.

Boelee, E., (2011). *Ecosystems for Water and Food Security, Nairobi: United Nations Environment Program*. International Water Management Institute, Colombo.

Buell, P. D., Anderson, E. N., De Pablo, M. M., & Oskenbay, M., (2020). Food by country. In: *Crossroads of Cuisine* (pp. 204–286). Brill.

Clasen, T., (2015). Household water treatment and safe storage to prevent diarrheal disease in developing countries. *Current Environmental Health Reports, 2*(1), 69–74.

Cohen, Y., (2021). *Advances in Water Desalination Technologies* (Vol. 17). World Scientific.

Cosgrove, C. E., & Cosgrove, W. J., (2012). *The United Nations World Water Development Report–N 4–The Dynamics of Global Water Futures: Driving Forces 2011–2050* (Vol. 2), UNESCO.

Dhanraj, S., Gopi, R., & Loganathan, G. B., (2019). An efficiency study on water extraction from air using thermophoresis method. In: *IOP Conference Series: Materials Science and Engineering* (Vol. 574, No. 1, p. 012003). IOP Publishing.

Dictionary, A. H., (2005). *The American Heritage Science Dictionary.*

Ebrahimi, A., Najafpour, G. D., & Kebria, D. Y., (2018). Performance of microbial desalination cell for salt removal and energy generation using different catholyte solutions. *Desalination, 432*, 1–9.

Fischetti, M., (2007). Fresh from the sea. *Scientific American, 297*(3), 118, 119.

Freeze, R. A., (1994). In: Cherry. J. A., (ed.), *Groundwater (1979)*. Printice Hall LNC. Englewood Cliffs, New Jersey.

Gosmartbricks.com/rain_water_harvesting_methods_in_individual (accessed on 24 February 2024).

Grizzetti, B., Liquete, C., Antunes, P., Carvalho, L., Geamănă, N., Giucă, R., & Woods, H., (2016). Ecosystem services for water policy: Insights across Europe. *Environmental Science & Policy, 66*, 179–190.

Hanley, S., (2017). *Inverted Umbrella Brings Clean Water & Clean Power to India*. Clean Technica.

Kinkade-Levario, H., (2007). *Design for Water: Rainwater Harvesting, Stormwater Catchment, and Alternate Water Reuse*. New SOCIETY PUBLISHERS.

MAJOR FLOODS RECHARGE AQUIFERS (2011). UNIVERSITY of New South Wales Science.

Network, R. W. S., (2017). *Rural Water Supply Network Self-Supply Site*. www.ruralwater-supply.net/en/self-supply (accessed on 24 February 2024).

Obaideen, K., Yousef, B. A., AlMallahi, M. N., Tan, Y. C., Mahmoud, M., Jaber, H., & Ramadan, M., (2022). An overview of smart irrigation systems using IoT. *Energy Nexus*, 100124.

Panagopoulos, A., Haralambous, K. J., & Loizidou, M., (2019). Desalination brine disposal methods and treatment technologies: A review. *Science of the Total Environment, 693*, 133545.

Patil, A. S., Patil, A. A., Patil, S. P., Saundade, S. D., Chugule, K. A., Katavare, S. B., & Panhalkar, S. S., (2021). *Analysis of the Agriculture Drought Severity and Spatial Extent Using Vegetation Health Index (VHI) in Manganga Watershed of Maharashtra*.

Pham, D., (2017). *New Rooftop Solar Hydro Panels Harvest Drinking Water and Energy at the Same Time*.

Postel, S., & Richter, B., (2012). *Rivers for Life: Managing Water for People and Nature*. Island press.

Shamsuddin, M. K. N., Sulaiman, W. N. A., Suratman, S., Zakaria, M. P., & Samuding, K., (2014). Conjunctive use of surface water and groundwater via the bank infiltration method. *Arabian Journal of Geosciences, 7*, 3731–3753.

Szatko, A., Tang, Z., Tian, S., & Rodie, S., (2012). *Gray or Green Infrastructure Approaches: Regulating Stormwater Management in Land Use Development*. LAND USE, 13.

Un.Org/waterforlifedecade/water_and_sustainable_development.schtml (accessed on 24 February 2024).

Watconman.org/introduction/Zibeline International Publishing.

Weerasekara, P., (2017). The United Nations world water development report 2017 wastewater. *Future of Food: Journal on Food, Agriculture and Society, 5*(2), 80, 81.

CHAPTER 11

Effect of ZnO Nanorods Treated by *Calycopteris floribunda* on the Growth of *Oryza sativa*: Green and Sustainable Chemistry

M. SOOSEN SAMUEL,[1] SHAN ABRAHAM,[1] JOHYNA ELS THOMAS,[1] ANNIE M. SHAJI,[1] and BINOY T. THOMAS[2]

[1]*Post-Graduate and Research Department of Physics, Catholicate College, Pathanamthitta, Kerala, India*

[2]*Post-Graduate and Research Department of Botany, Catholicate College, Pathanamthitta, Kerala, India*

ABSTRACT

The progress of green chemistry in the synthesis of ZnO nanorods using plant extract has garnered significant attention. This study presents the utilization of vine water from *Calycopteris floribunda* Lam for the hydrothermal synthesis of ZnO nanorods. *Calycopteris floribunda* is a large climbing shrub known for its diverse medicinal properties found in its stem and leaves. This research focuses on the synthesis and characterization of ZnO nanorods using a green chemical method. The X-ray diffraction (XRD) pattern reveals a rod-like morphology, while UV-Vis absorption displays the characteristic absorption peak of ZnO. Photoluminescence (PL) highlights its emission properties. The growth rate of *Oryza sativa* is controlled by ZnO nanorods.

Sustainable Environment and Health: Practical Strategies. Leonardo Sepúlveda Torre, Ann Rose Abraham, Neha Kanwar Rawat, & A. K. Haghi (Eds.)

11.1 INTRODUCTION

In recent years, nanotechnology has become one of the most important and exciting forefront fields in physics, chemistry, engineering, and biology (Bharat, 2004; Jan & Andreas, 2002; Rainer, 2008; Whitesides, 2003; Bogunia-Kubik & Sugisaka, 2002; Alexandra, 2003; Laura, 2003; Andreas, 2011). Nanotechnology involves the manipulation of matter on an atomic, molecular, and supramolecular scale. The main purpose of developing nanotechnology is to enhance our understanding of nature at the molecular level, increase productivity, improve healthcare, and push the boundaries of sustainable development and human potential (Garcia-Martinez, 2010; Mohamed, 1998; Mark & Jean-Yves, 2007; Geller, Hopfer, & Bimberg, 2008).

The synthesis of zinc oxide nanoparticles is particularly important due to its significant applications in the biomedical field (Lloyd, 2010; Daniel, 2006; Sahoo, Parveen, & Panda, 2007; Mohamadi et al., 2006; Prasad et al., 2012). While chemical and green methods are used for nanomaterial synthesis, the green approach is a better choice as it is eco-friendly (Mohamadi et al., 2006; Prasad et al., 2012; Cakmak, 2000). The green synthesis procedure offers advantages over physiochemical methods, such as simplicity, cost-effectiveness, good stability of nanoparticles, shorter time consumption, non-toxic by-products, and the ability for large-scale synthesis (Mohamadi et al., 2006; Prasad et al., 2012; Cakmak, 2000). *Calycopteris floribunda* Lam. is a large climbing scandent shrub that can grow up to 5–10 m long. It has slender, brown-streaked diffuse vines that occasionally twine around supports and store water abundantly. This plant is commonly found in the low-lying tropical evergreen forests of central and southern parts of India. The stem and leaves are said to have medicinal properties (Cakmak, 2000).

The progress of green chemistry in the synthesis of ZnO nanorods using plant extract has garnered great attention. This study reports the use of vine water from *Calycopteris floribunda* (Roxb.) Lam for the synthesis of ZnO Nanorods by the hydrothermal method. *Calycopteris floribunda* is a large climbing shrub, and its stem and leaves are said to have significant medicinal properties. The present study describes the synthesis and characterization of ZnO Nanorods prepared using a green chemical method and investigates its antitoxic effect on the growth of *Oryza sativa*.

11.2 MATERIALS AND METHODS

11.2.1 GREEN APPROACH IN THE SYNTHESIS OF ZNO NANORODS BY HYDROTHERMAL METHOD

The green approach in the synthesis of ZnO Nanorods was achieved by treating it with the vine water of *Calycopteris floribunda*. The alkali solution of Zn was prepared by mixing 0.1 molar zinc chloride ($ZnCl_2$) in 500 ml of distilled water. $ZnCl_2$ is highly soluble in water and forms a clear solution. About 1 ml of vine water of *Calycopteris floribunda* was added to the solution, which was then thoroughly mixed using a magnetic stirrer. A small amount of ammonium hydroxide (NH_4OH) was added drop by drop. The pH of the solution was noted simultaneously. When the pH reached 12 by adding a sufficient amount of NH_4OH, the solution was transferred to an autoclave flask. The flask was then placed in an oven preheated at 100°C for about 90 minutes. After 90 minutes, the oven was turned off. Carefully, the sample was transferred from the flask to a beaker and the solution was thoroughly washed for several days with deionized water until the pH reached 7. Then, the solution was carefully filtered without losing the precipitate. The precipitate was dried in air and powdered for further characterizations (Soosen, Lekshmi, & George, 2009).

The above procedure is repeated for different concentrations of vine water of *Calycopteris floribunda*. Instead of adding 1 ml, concentrations of 5 ml, 15 ml, 10 ml, 25 ml, and 50 ml are used. Additionally, one sample of pure Nanorods is synthesized without adding vine water of *Calycopteris floribunda*. As a result, we obtain seven samples: pure ZnO Nanorods, ZnO Nanorods treated with 1 ml, 5 ml, 15 ml, 10 ml, 25 ml, and 50 ml vine water of *Calycopteris floribunda*.

The chemical reactions are as follows:

$$ZnCl_2 \rightarrow Zn^{2+} + 2Cl^-$$

$$2NH_3.2H_2O \rightarrow 2NH_4^+ + 2OH^-$$

$$Zn^{2+} + 4NH_3.H_2O \rightarrow Zn\,(NH_3)_4^{2+} + H_2O$$

$$Zn\,(NH_3)_4^{2+} + 2OH^- \rightarrow ZnO + 4NH_3 + H_2O$$

The overall reaction is:

$$ZnCl_2 + 6NH_3.3H_2O \rightarrow ZnO + 4NH_3 + 2NH_4Cl + 2H_2O$$

11.2.2 STUDY OF ANTITOXIC EFFECT – GROWTH OF ORYZA SATIVA

In order to study the antitoxic effect of ZnO, we applied the prepared sample of ZnO Nanorods synthesized chemically and by a green approach (Nanorods treated with vine water of *Calycopteris floribunda*) on plants (Shah & Belozerova, 2009; Pramod, Dhoke, & Khanna, n.d.; Cakmak, 2000; Chopra, Nayar, & Chopra, 1956). *Oryza sativa* (Rice plant) was used for the toxicological studies. To apply it on plants, different ppm (parts per million) solutions were prepared. Solutions were prepared by adding bulk ZnO, ZnO Nanorods synthesized chemically, which were treated with 1 ml, 5 ml, 15 ml, 25 ml, and 50 ml of vine water of *Calycopteris floribunda* to distilled water with concentrations of 200 ppm, 600 ppm, 800 ppm, 1,000 ppm, and 2,000 ppm. Thus, 35 samples of ppm solution were prepared and kept in different bottles, labeled, respectively as shown in Table 11.1. A control solution of distilled water was also used.

TABLE 11.1 Sample Solution Labeled Names

Sample Solution Names	200 ppm	600 ppm	800 ppm	1,000 ppm	2,000 ppm
Bulk ZnO	B1	B2	B3	B4	B5
ZnO nanorods	C1	C2	C3	C4	C5
ZnO + 1 ml vine water	P1	P2	P3	P4	P5
ZnO + 5 ml vine water	P6	P7	P8	P9	P10
ZnO + 15 ml vine water	P11	P12	P13	P14	P15
ZnO + 25 ml vine water	P16	P17	P18	P19	P20
ZnO + 50 ml vine water	P21	P22	P23	P24	P25
Control	C				

Seeds of *Oryza sativa* (rice plant) were purchased locally. The seeds were immersed in each sample solution for 24 hours for germination. Garden soil was autoclaved for thorough sterilization. The autoclaved soil was then placed in different plastic cups for growing the seeds. Each seed sample was sown in a separate cup. The treatment was conducted with three replicates and labeled accordingly. The remaining sample solution was used to water the corresponding seeds daily. Starting from the third day, the germination percentage, root length, and shoot length were measured. The differences in growth patterns were measured and the results were presented as an average. The germination percentage, root length, and shoot length were statistically analyzed using Excel software.

11.3 RESULT AND DISCUSSION

11.3.1 SCANNING ELECTRON MICROSCOPY (SEM) IMAGES

The morphology of ZnO was analyzed using scanning electron microscopy (SEM). Figure 11.1 shows the SEM image of ZnO Nanorods synthesized via the hydrothermal method. Figure 11.2 displays the SEM image of ZnO Nanorods treated with vine water of *Calycopteris floribunda*.

FIGURE 11.1 *SEM image of ZnO nanorods.*

In general, the morphology of ZnO nanorods remains the same after treatment with vine water of *Calycopteris floribunda* (Shindo & Hiraga, 1998).

FIGURE 11.2 SEM image of ZnO nanorods treated with 25 ml vine water of *Calycopteris floribunda.*

11.4 X-RAY DIFFRACTION (XRD) PATTERN ANALYSIS

The internal structure of a crystal is primarily studied using XRD methods. Figure 11.3 displays the XRD pattern of ZnO Nanorods. Figures 11.4–11.9 exhibit the XRD pattern of ZnO Nanorods treated with different concentrations of water vine from *Calycopteris floribunda*. By comparing the 2θ and d values of each sample with the JCPDS values of ZnO, it is evident that the samples are ZnO Nanorods. The presence of strong and distinct diffraction peaks at 2θ = 31.7° (hkl =100, d = 2.8143), 34.4° (hkl = 002, d = 2.6032), 36.2° (hkl = 101, d = 2.4759), 47.5° (hkl = 102, d = 1.9111), and 56.6° (hkl = 110, d = 1.6247) planes, respectively indicates the formation of ZnO nanorods (Tables 11.2–11.8).

The average crystalline size of the particles in each sample can be calculated using Scherrer's formula.

$$L = k\,\lambda/\beta\cos\theta$$

where; 'k' is the dimensionless shape factor, with a value close to unity. The shape factor typically has a value of about 0.9, but it varies with the actual shape of the crystallite; 'λ' = 1.5418 Å, is the X-ray wavelength; 'β' is the line broadening at half the maximum intensity (full width at half maximum (FWHM)), after subtracting the instrumental line broadening, in radians. This quantity is also sometimes denoted as Δ(2θ); and θ is the Bragg angle.

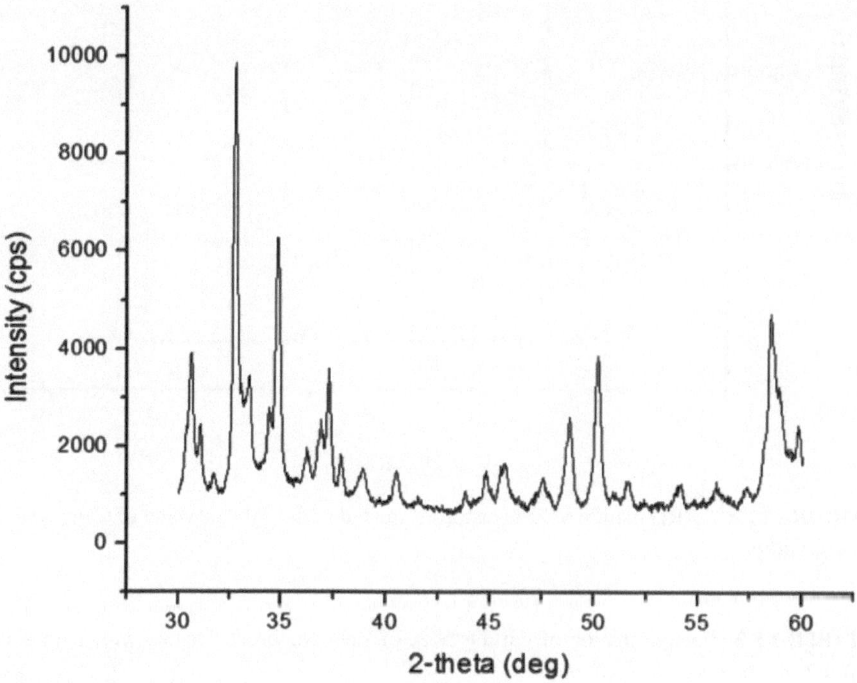

FIGURE 11.3 XRD pattern of ZnO nanorods.

TABLE 11.2 Determination of Particle Size of ZnO Nanorods

Observed 2θ (Degree)	θ (Radian)	FWHM (Degree)	β (Radian)	Particle Size (nm)
32.776	0.286024316	0.213	0.003717548	38.90687813
34.806	0.303739393	0.214	0.003735001	38.93400691
36.223	0.31610504	0.276	0.004817105	30.3077448
47.54	0.414864413	0.33	0.005759582	26.32554496
55.85	0.487382782	0.4	0.006981311	22.49555656

Note: The average particle size = 31.39 nm.

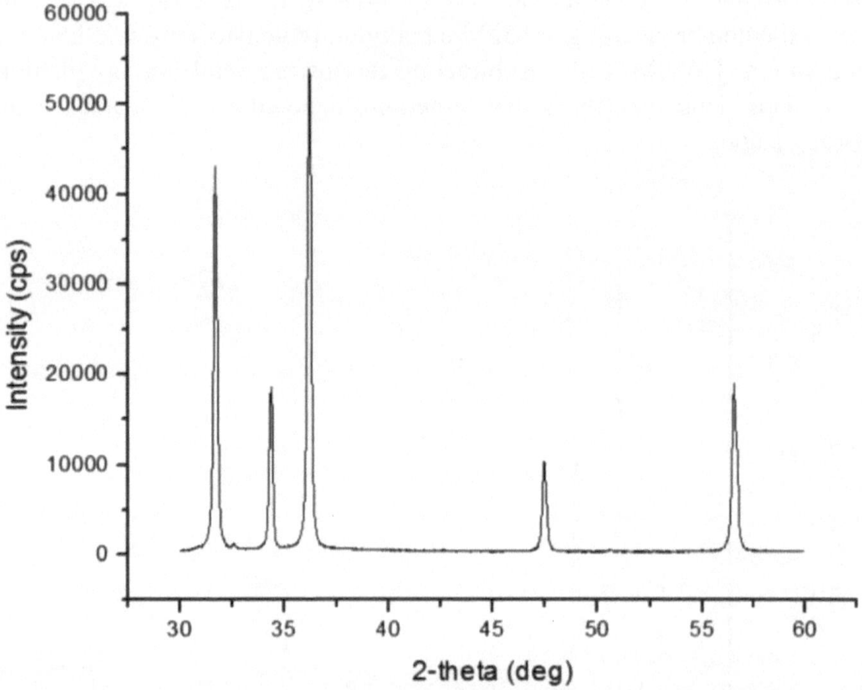

FIGURE 11.4 XRD pattern of ZnO nanorods treated with 1 ml vine water of *Calycopteris floribunda*.

TABLE 11.3 Determination of Particle Size of ZnO Nanorods Treated with 1 ml Vine Water of *Calycopteris floribunda*

Observed 2θ (Degree)	θ (Radian)	FWHM (Degree)	β (Radian)	Particle Size (nm)
31.6851	0.276504426	0.1806	0.003152062	45.7608092
34.339	0.299664053	0.1632	0.002848375	50.98846342
36.171	0.315651255	0.1844	0.003218384	45.35627484
47.435	0.413948116	0.181	0.003159043	47.97750818
56.494	0.493002737	0.206	0.003595375	43.81188823

Note: The average particle size = 46.78 nm.

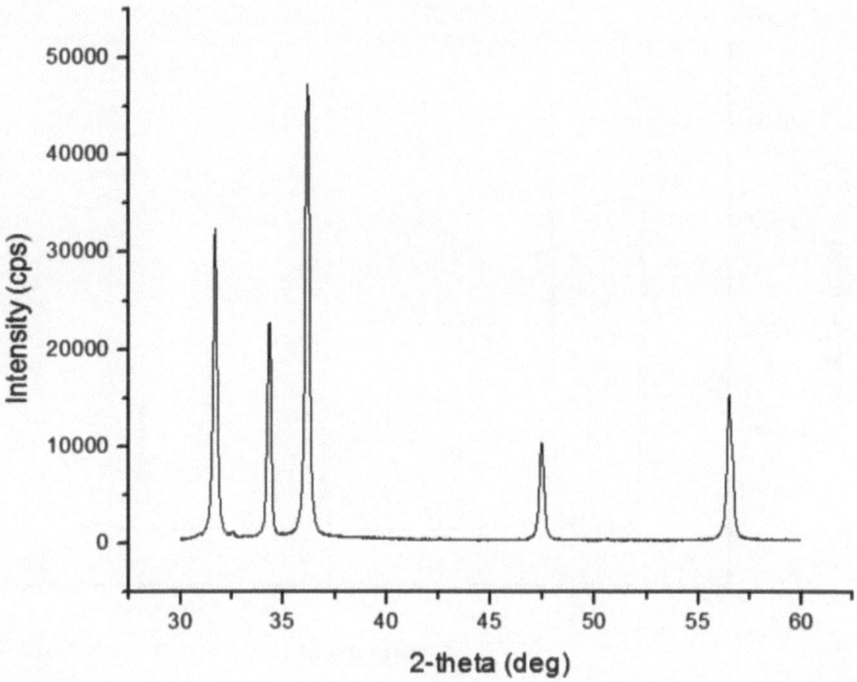

FIGURE 11.5 XRD pattern of ZnO nanorods treated with 5 ml vine water of *Calycopteris floribunda*.

TABLE 11.4 Determination of Particle Size of ZnO Nanorods Treated with 5 ml Water Vine of *Calycopteris floribunda*

Observed 2θ (Degree)	θ (Radian)	FWHM (Degree)	β (Radian)	Particle Size (nm)
31.669	0.276363927	0.2076	0.0036233	39.80767214
34.327	0.299559333	0.172	0.003001964	48.37818632
36.148	0.315450543	0.204	0.003560469	40.99582878
47.426	0.413869576	0.212	0.003700095	40.96051593
56.463	0.492732212	0.224	0.003909534	40.28543644

Note: The average particle size = 42.09 nm.

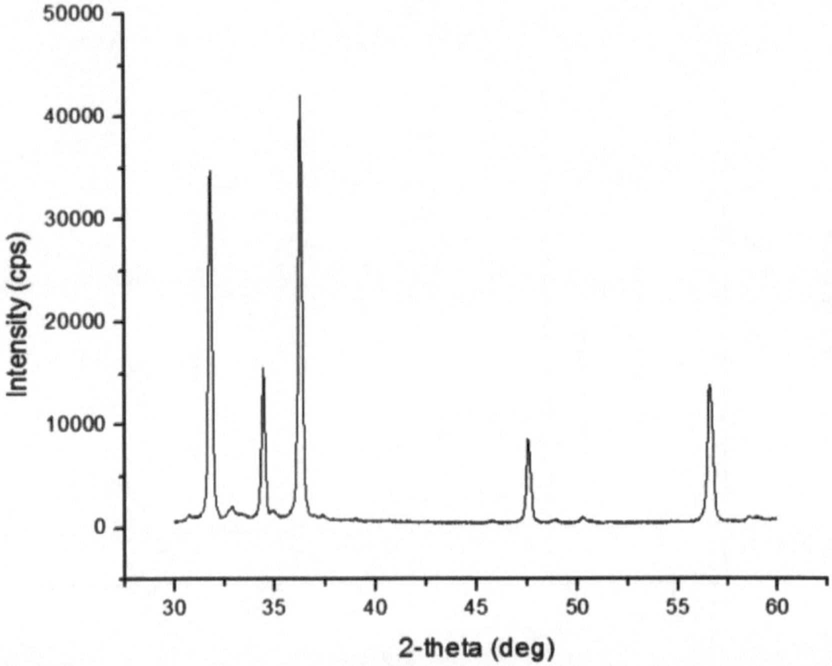

FIGURE 11.6 XRD pattern of ZnO nanorods treated with 10 ml vine water of *Calycopteris floribunda.*

TABLE 11.5 Determination of Particle Size of ZnO Nanorods Treated with 10 ml Vine Water of *Calycopteris floribunda*

Observed 2θ (Degree)	θ (Radian)	FWHM (Degree)	β (Radian)	Particle Size (nm)
31.789	0.277411124	0.1911	0.003335321	43.25762685
34.439	0.300536717	0.185	0.003228856	44.99224132
36.268	0.316497739	0.192	0.003351029	43.57298306
47.516	0.414654973	0.194	0.003385936	44.77643685
56.562	0.493596149	0.231	0.004031707	39.0828088

Note: The average particle size = 43.14 nm.

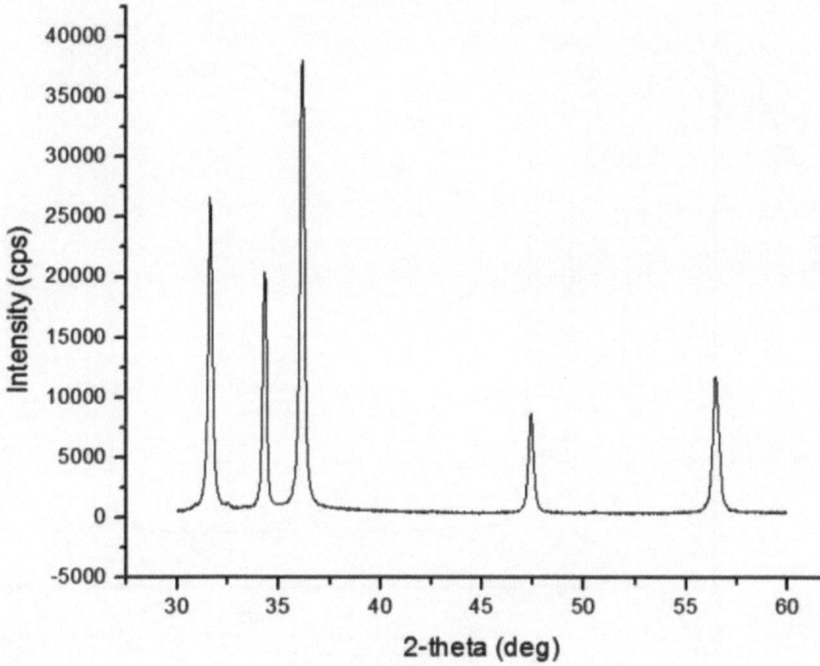

FIGURE 11.7 XRD pattern of ZnO nanorods treated with 15 ml vine water of *Calycopteris floribunda.*

TABLE 11.6 Determination of Particle Size of ZnO Nanorods Treated with 15 ml Vine Water of *Calycopteris floribunda*

Observed 2θ (Degree)	θ (Radian)	FWHM (Degree)	β (Radian)	Particle Size (nm)
31.641	0.276119581	0.226	0.003944441	36.56416065
34.301	0.299332441	0.194	0.003385936	42.88899584
36.129	0.315284736	0.234	0.004084067	35.73801999
47.39	0.413555417	0.241	0.00420624	36.02668758
56.417	0.492330786	0.275	0.004799651	32.80725095

Note: The average particle size = 36.81 nm.

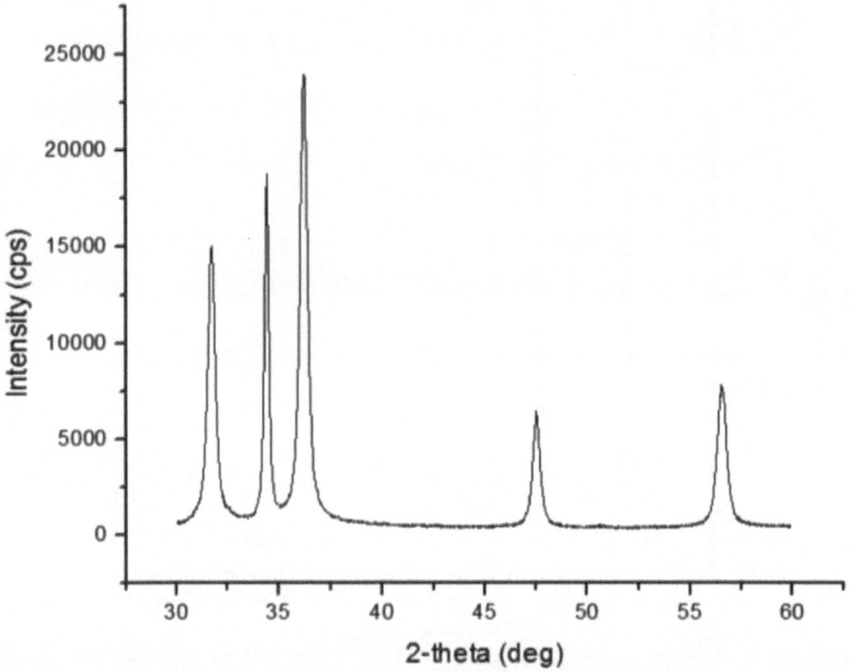

FIGURE 11.8 XRD pattern of ZnO nanorods treated with 25 ml vine water of *Calycopteris floribunda.*

TABLE 11.7 Determination of Particle Size of ZnO Nanorods Treated with 25 ml Vine Water of *Calycopteris floribunda*

Observed 2θ (Degree)	θ (Radian)	FWHM (Degree)	β (Radian)	Particle Size (nm)
31.712	0.276739172	0.422	0.007365283	19.58519653
34.406	0.300248738	0.222	0.003874628	37.49018992
36.216	0.316043954	0.404	0.007051124	20.70487736
47.47	0.414253548	0.374	0.006527526	23.22217863
56.505	0.49309873	0.496	0.008656826	18.19700503

Note: The average particle size = 23.84 nm.

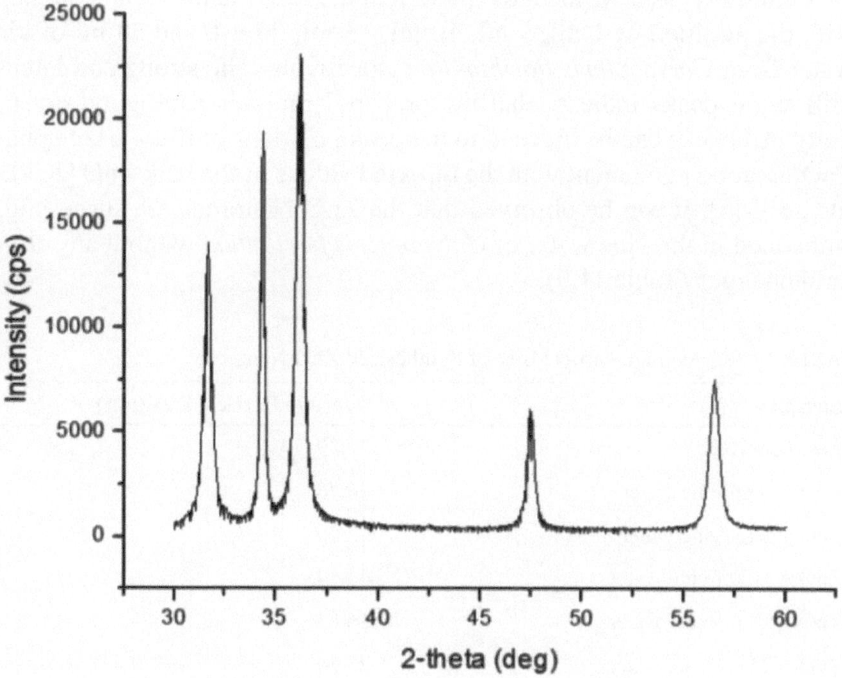

FIGURE 11.9 XRD pattern of ZnO nanorods treated with 50 ml vine water of *Calycopteris floribunda.*

TABLE 11.8 Determination of Particle Size of ZnO Nanorods Treated with 50 ml Vine Water of *Calycopteris floribunda*

Observed 2θ (Degree)	θ (Radian)	FWHM (Degree)	β (Radian)	Particle Size (nm)
31.575	0.275543623	0.43	0.007504909	19.21430994
34.338	0.299655326	0.161	0.002809978	51.68506083
36.211	0.316000321	0.16	0.002792524	52.27906946
47.248	0.412316234	0.32	0.005585049	27.1178723
56.485	0.492924198	0.542	0.009459677	16.65104833

Note: The average particle size = 33.39 nm.

Figures 11.9 shows the XRD pattern of the ZnO Nanorods and the same with the addition of 1 ml, 5 ml, 10 ml, 15 ml, 25 ml, and 50 ml of vine water from *Calycopteris floribunda*, respectively. The strong and intense diffraction peaks indicate that the prepared samples have good crystallinity. All peaks can be indexed to the peak positions of the wurtzite phase ZnO, in good agreement with the reported values of the bulk ZnO [JCPDS file 36–145]. It can be observed that the ZnO Nanorods are successfully embedded in the vine water of *Calycopteris floribunda* without any structural changes (Table 11.9).

TABLE 11.9 Average Particle Size of Synthesized ZnO Nanorods

Sample	Average Particle Size (nm)
ZnO nanorods	31.39
ZnO + 1 ml vine water	46.78
ZnO + 5 ml vine water	42.09
ZnO + 10 ml vine water	43.14
ZnO + 15 ml vine water	36.81
ZnO + 25 ml vine water	23.84
ZnO + 50 ml vine water	33.39

11.5 UV-VISIBLE ABSORPTION SPECTRA ANALYSIS

Figures 11.10–11.16 show absorption spectrum of ZnO Nanorods and it treated with vine water of *Calycopteris floribunda*.

Figures 11.10–11.16 show the absorption spectrum of ZnO Nanorods. It was obtained near 377 nm for ZnO Nanorods and treated with vine water of *Calycopteris floribunda*. Spectral lines observed in the UV-Visible region correspond to the energy difference between two well-defined electronic energy levels of the absorbing atom or molecule. When an atom or molecule absorbs energy in that region, its electrons are promoted from a state of lower energy, which is the ground state, to orbitals of higher energy (Table 11.10).

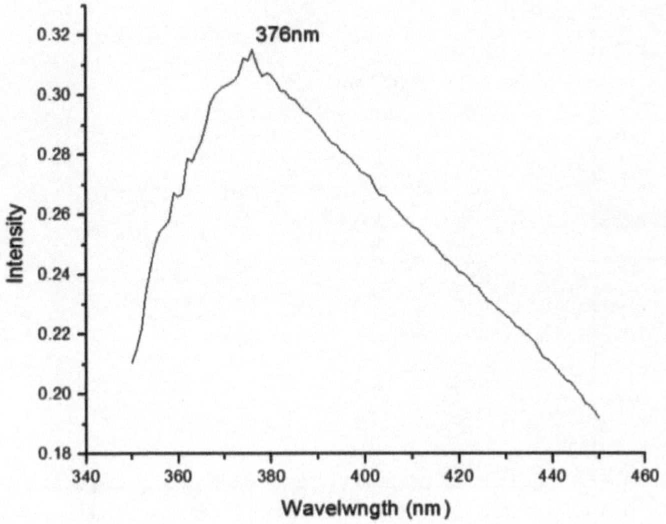

FIGURE 11.10 Absorption spectrum of ZnO nanorods.

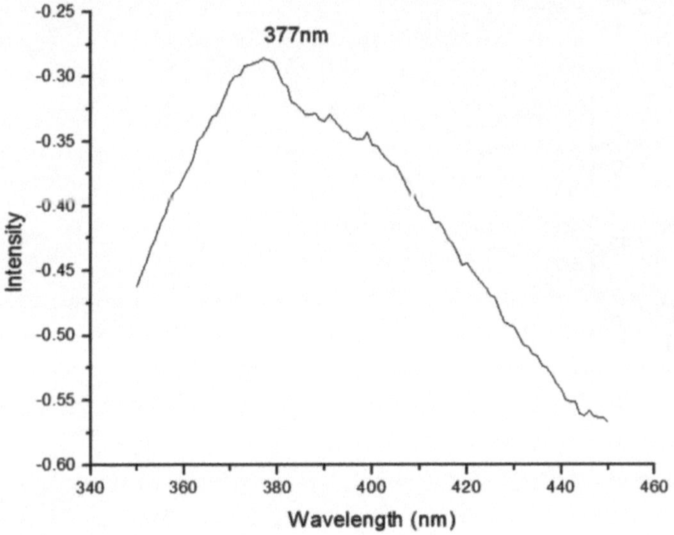

FIGURE 11.11 Absorption spectrum of ZnO nanorods treated with 1 ml vine water of *Calycopteris floribunda.*

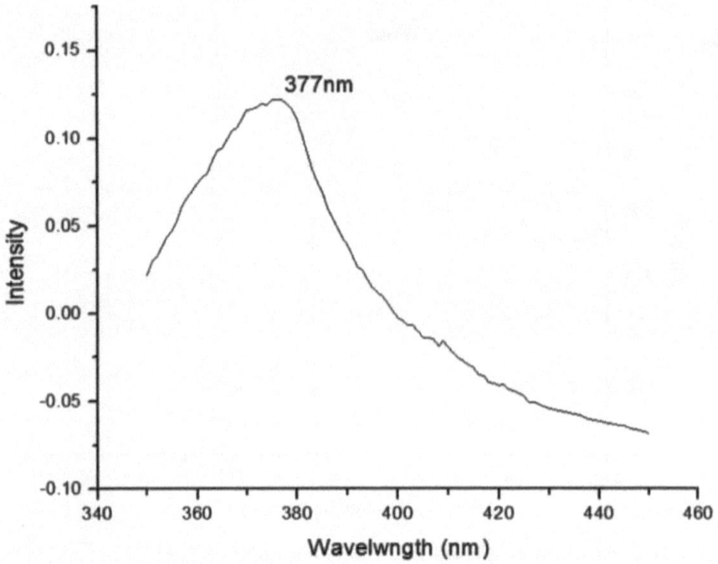

FIGURE 11.12 Absorption spectrum of ZnO nanorods treated with 5 ml vine water of *Calycopteris floribunda.*

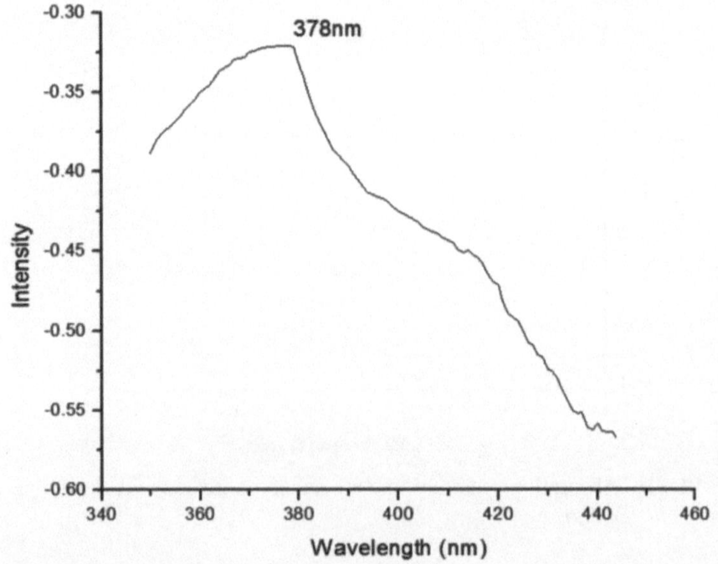

FIGURE 11.13 Absorption spectrum of ZnO nanorods treated with 10 ml vine water extract of *Calycopteris floribunda.*

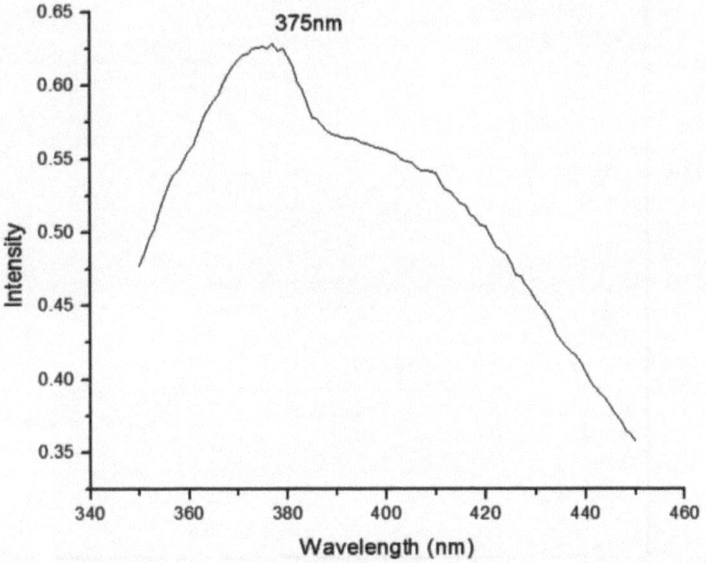

FIGURE 11.14 Absorption spectrum of ZnO nanorods treated with 15 ml vine water of *Calycopteris floribunda.*

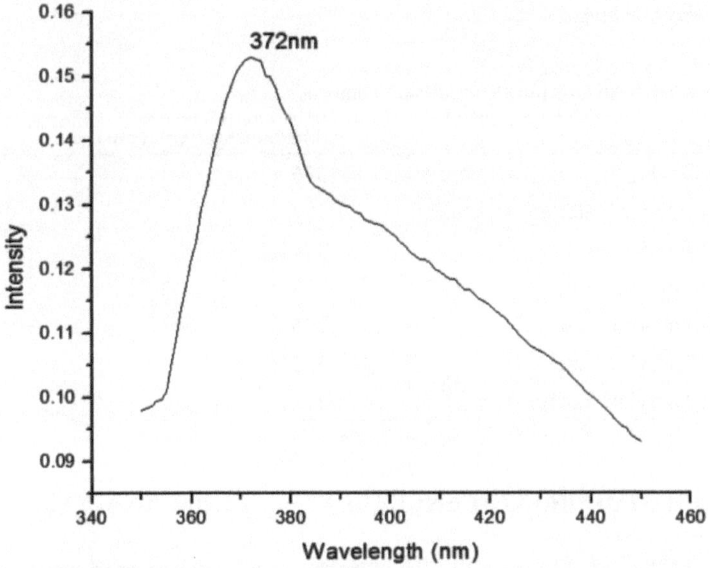

FIGURE 11.15 Absorption spectrum of ZnO nanorods treated with 25 ml vine water of *Calycopteris floribunda.*

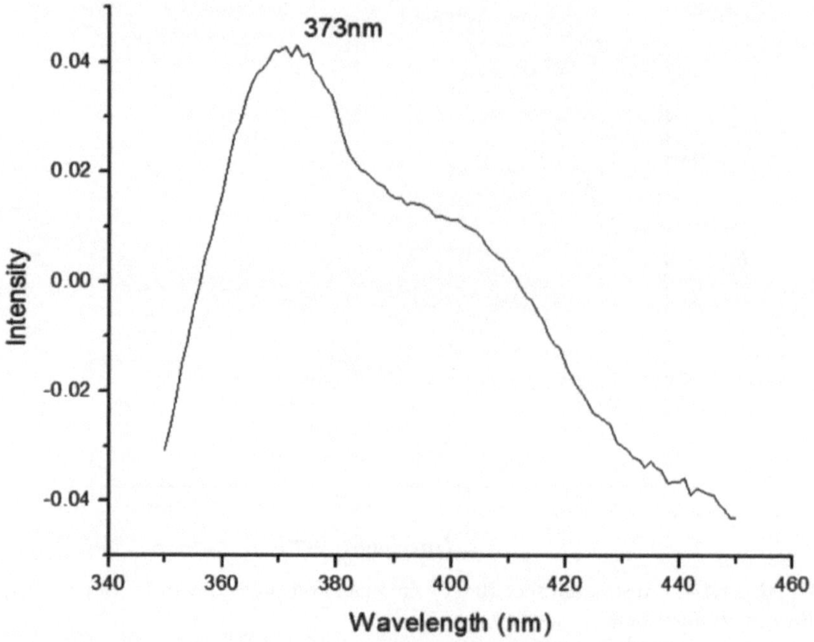

FIGURE 11.16 Absorption spectrum of ZnO nanorods treated with 50 ml vine water of *Calycopteris floribunda.*

TABLE 11.10 Absorption Peak of ZnO Nanorods

Sample	Absorption Peak (nm)
ZnO nanorods	376
ZnO + 1 ml vine water	377
ZnO + 5 ml vine water	377
ZnO + 10 ml vine water	378
ZnO + 15 ml vine water	375
ZnO + 25 ml vine water	372
ZnO + 50 ml vine water	373

11.6 PHOTOLUMINESCENCE (PL) SPECTRUM ANALYSIS

Figures 11.17–11.23 show the PL spectrum of ZnO nanorods and it treated with vine water of *Calycopteris floribunda.*

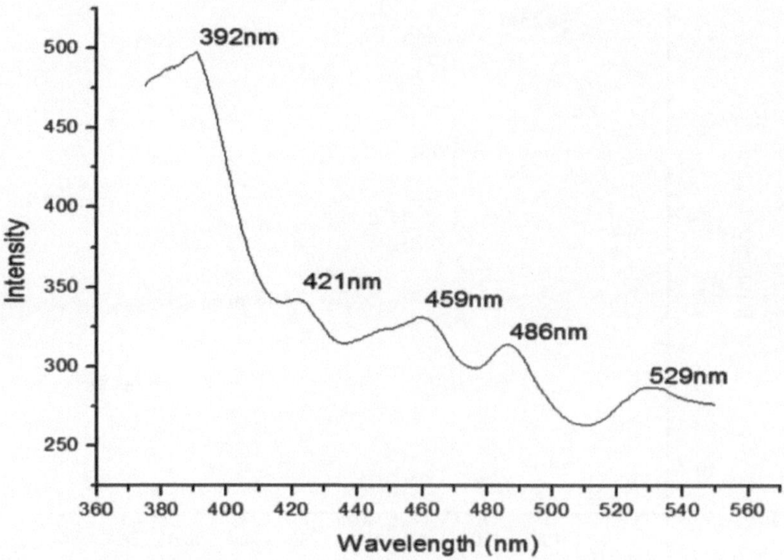

FIGURE 11.17 PL spectrum of ZnO nanorods.

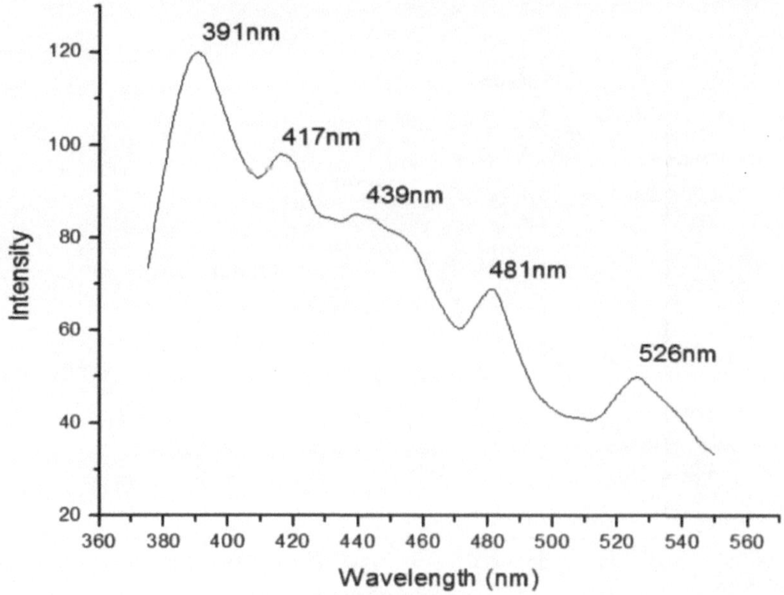

FIGURE 11.18 PL spectrum of ZnO nanorods treated with 1 ml vine water of *Calycopteris floribunda*.

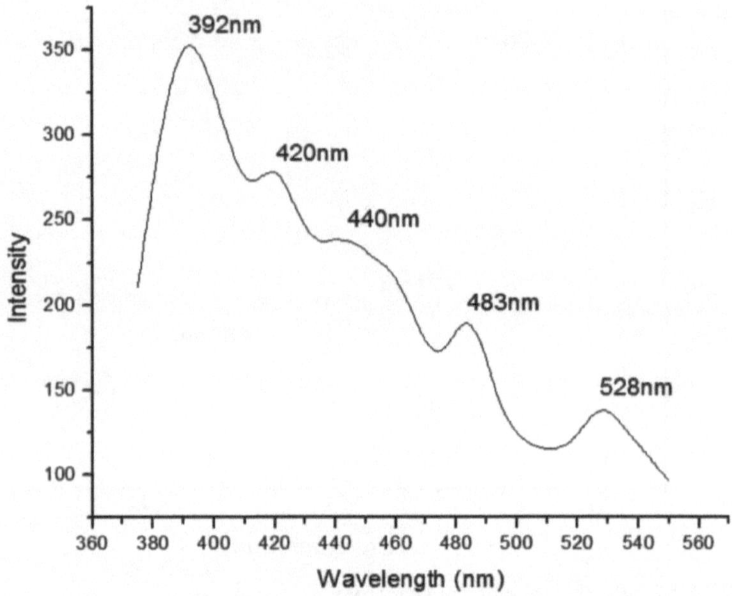

FIGURE 11.19 PL spectrum of ZnO nanorods treated with 5 ml vine water of *Calycopteris floribunda.*

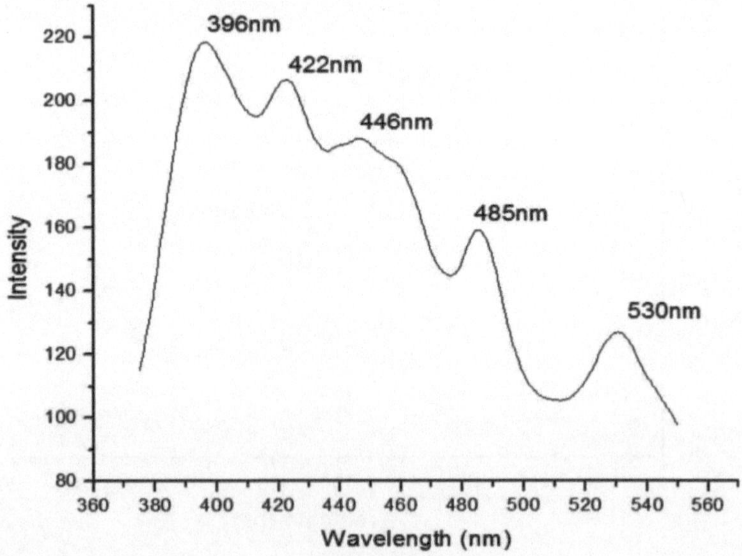

FIGURE 11.20 PL spectrum of ZnO nanorods treated with 10 ml vine water of *Calycopteris floribunda.*

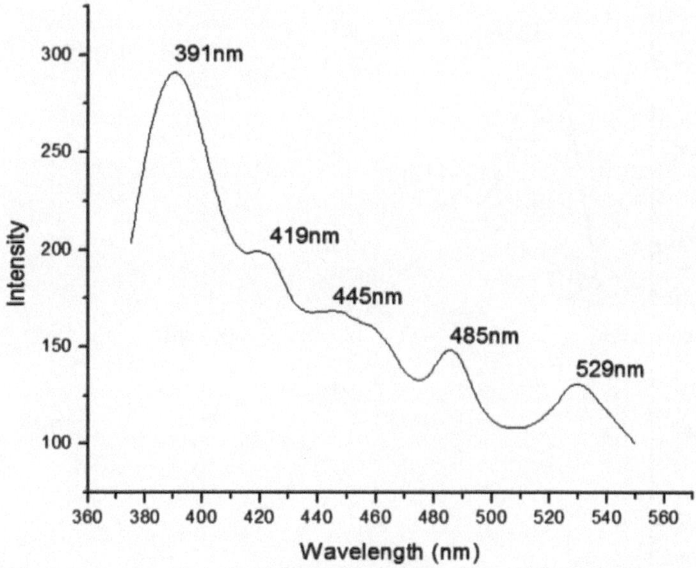

FIGURE 11.21 PL spectrum of ZnO nanorods treated with 15 ml vine water of *Calycopteris floribunda.*

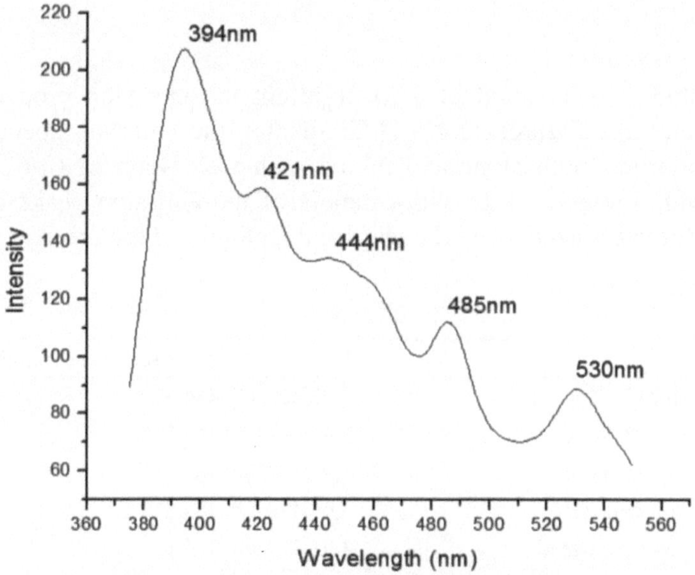

FIGURE 11.22 PL spectrum of ZnO nanorods treated with 25 ml vine water of *Calycopteris floribunda.*

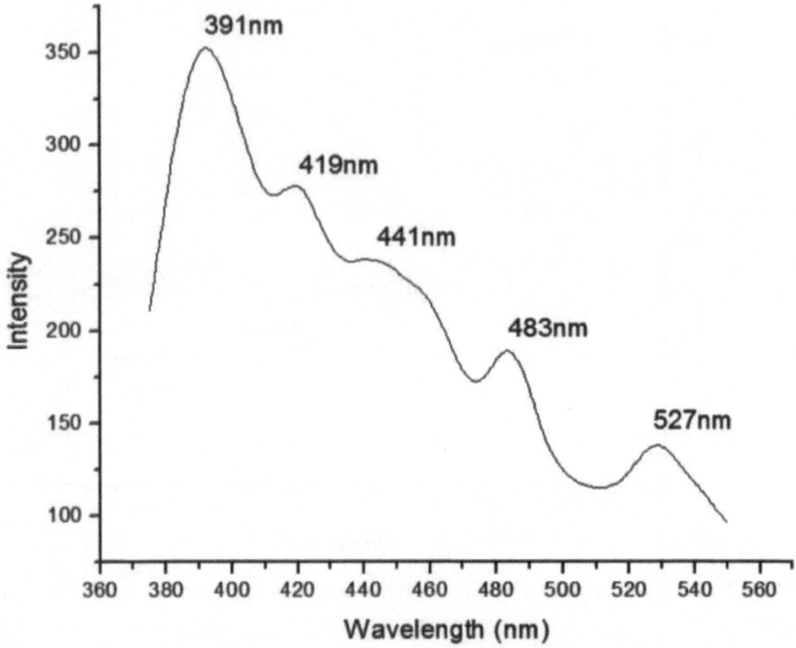

FIGURE 11.23 PL spectrum of ZnO nanorods treated with 50 ml vine water of *Calycopteris floribunda.*

PL studies were conducted to highlight the emission properties of ZnO Nanorods. Figures 11.17–11.23 display the emission spectrum of ZnO Nanorods, both alone and mixed with vine water of *Calycopteris floribunda.* Table 11.11 provides details of the emission peaks of ZnO Nanorods synthesized using the chemical method and the green approach.

TABLE 11.11 Emission Peaks of ZnO Nanorods

Sample	Emission Peaks at the Wave Length (nm)
ZnO nanorods	3,92,421,459,486,529
ZnO + 1 ml vine water	3,91,417,439,481,526
ZnO + 5 ml vine water	3,92,420,440,483,528
ZnO + 10 ml vine water	3,96,422,446,485,530
ZnO + 15 ml vine water	3,91,419,445,485,529
ZnO + 25 ml vine water	3,94,421,444,485,530
ZnO + 50 ml vine water	3,91,419,441,483,527

All the emission peaks of the samples were observed in the UV-visible region. The UV emission line was observed at wavelengths near 392 nm, 419 nm, 440 nm, 483 nm, and 529 nm, respectively.

The blue emission corresponds to the electron transition from the level of Zinc interstitial (Zni) to the valence band. The energy interval from the level of interstitial Zn to the valence band is exactly consistent with the energy of the blue emission observed in our experiment. The figures also show that the green emission is very weak compared to the violet-blue emission. These results indicate that the as-prepared ZnO Nanorods have few oxygen vacancies as structural defects but more zinc interstitials. Hence, the ZnO Nanorods have more potential in the violet-blue region.

It is also important to note that there is no change in the emission peaks of ZnO Nanorods when treated with vine water of *Calycopteris floribunda* at different concentrations.

11.7 ANTITOXIC EFFECT OF ZNO NANORODS–STATISTICAL ANALYSIS

The antitoxic effect of ZnO Nanorods was studied by applying them to plants. The experimental plant used was *Oryza sativa* (rice plant). The selected seeds were treated with different ppm solutions, and the shoot and root growth were observed at regular intervals (Figure 11.24).

It was found that the presence of bulk zinc oxide in the experiment promoted growth, enhancing the shoot growth of rice at different ppm solutions. The maximum growth was observed at 800 ppm solution. At 2,000 ppm, the growth of the shoot slightly slows down compared to that of other ppm solutions (Figure 11.25).

In the case of root elongation, a bulk amount of ZnO promotes the elongation of roots in *Oryza sativa*. Significant growth of the root was also observed in higher concentrations (Figure 11.26).

Chemically synthesized pure ZnO nanorods have accelerated the shoot growth of *Oryza sativa* at different ppm solutions. The maximum growth of the shoot was observed at 800 ppm (Figure 11.27).

Compared to bulk, chemically synthesized pure ZnO nanorods had a more influential effect on the growth of *Oryza sativa*. The same effect was also observed in root elongation (Figure 11.28).

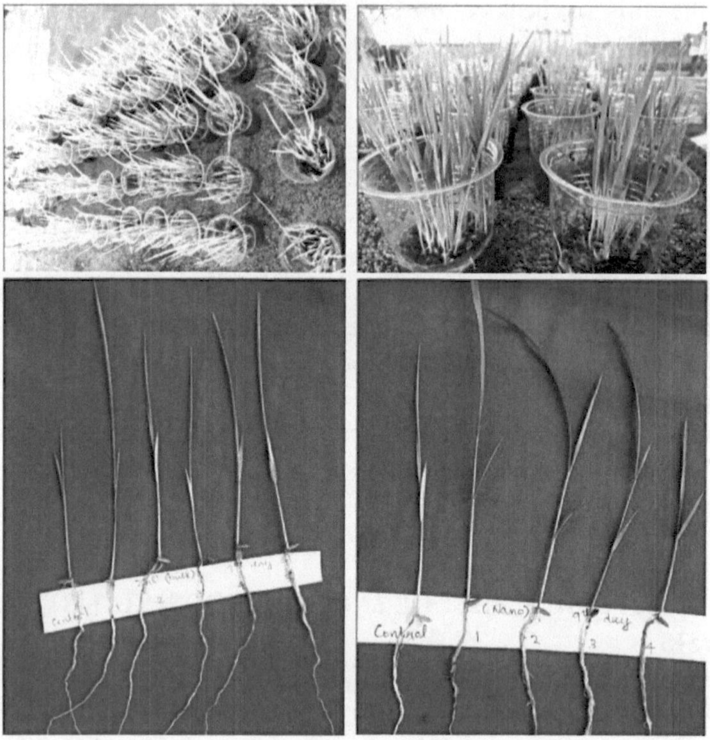

FIGURE 11.24 Growth rate analysis of *Oryza sativa*.

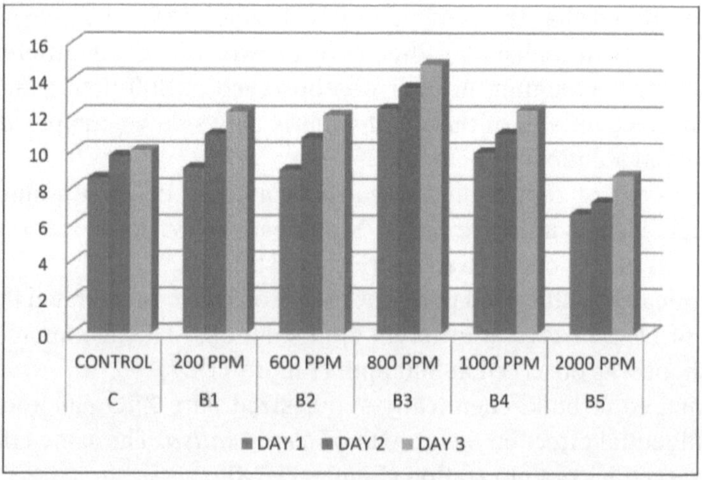

FIGURE 11.25 Shoot growth of *Oryza sativa* added with bulk ZnO.

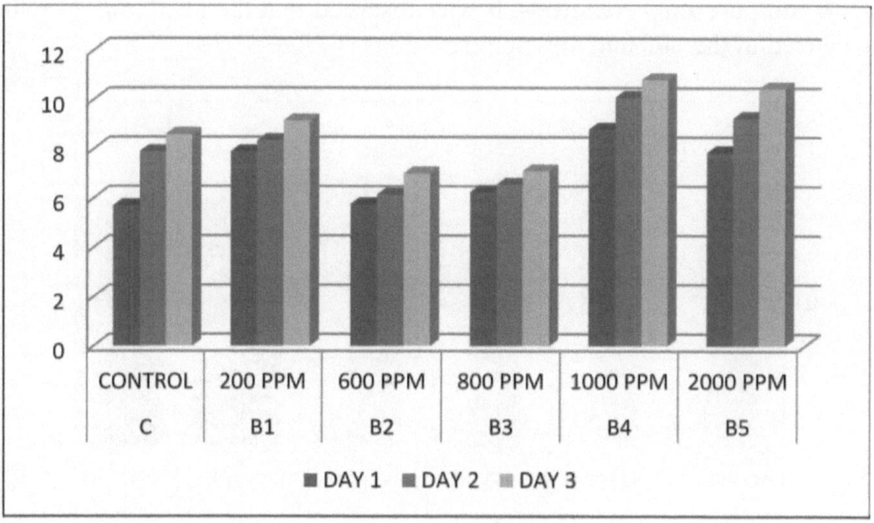

FIGURE 11.26 Root growth of *Oryza sativa* added with bulk ZnO.

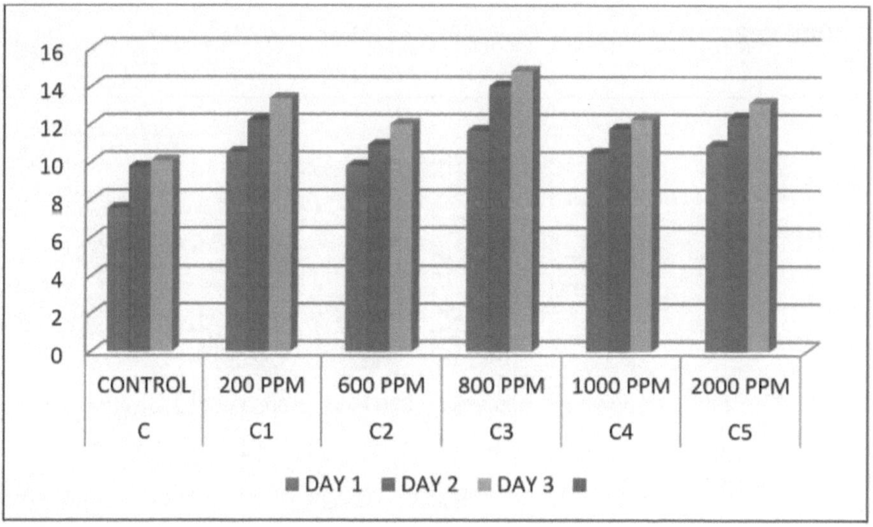

FIGURE 11.27 Shoot growth of *Oryza sativa* added with chemically synthesized pure ZnO nanorods.

ZnO Nanorods treated with *Calycopteris floribunda* in 1 ml of vine water exhibited a greater accelerating effect on shoot growth, in comparison to bulk and pure ZnO Nanorods. It was observed that the maximum growth occurred in the 800 and 600 ppm solutions (Figure 11.29).

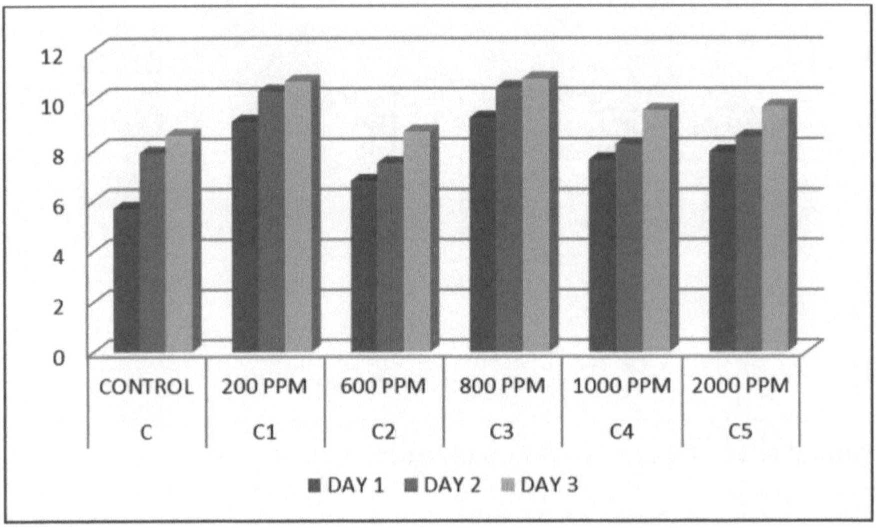

FIGURE 11.28 Root growth of *Oryza sativa* added with pure ZnO nanorods.

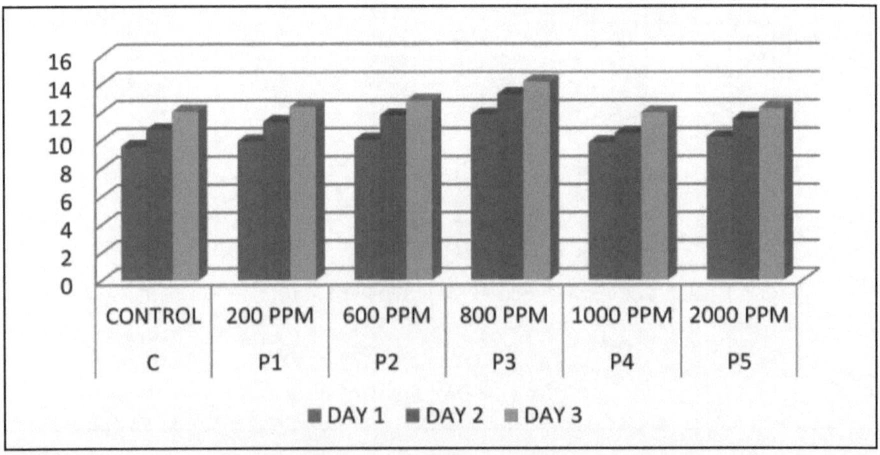

FIGURE 11.29 Shoot growth of *Oryza sativa* added with ZnO nanorods + 1 ml vine water.

The growth of the experimental plant increased with the addition of ZnO nanorods treated with *Calycopteris floribunda*. It was higher for the 800-ppm solution (Figure 11.30).

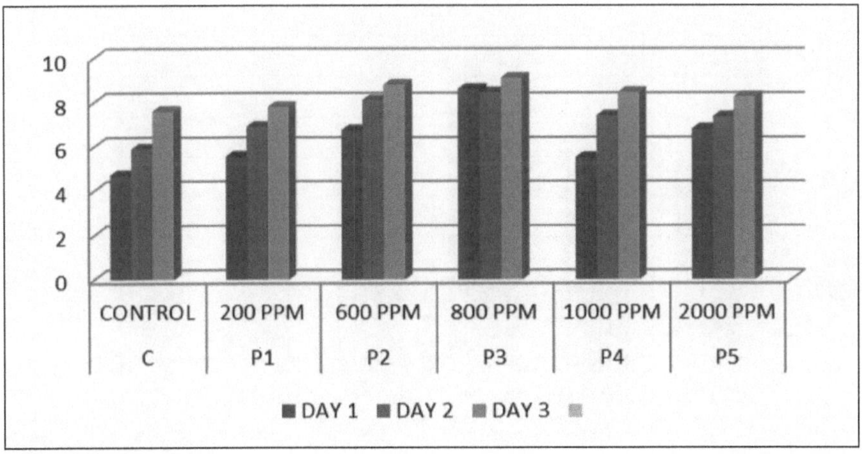

FIGURE 11.30 Root growth of *Oryza sativa* added with nanorods of ZnO + 1 ml vine water.

As observed from 200 ppm to 2,000 ppm solutions of ZnO Nanorods treated with *Calycopteris floribunda* at 5-, 15-, 25-, and 50-ml vine water, it was witnessed that the shoot growth of rice plants almost steadily increased. In the control group, growth was still lower, but at 200 ppm, maximum growth was observed (Figures 11.31–11.34).

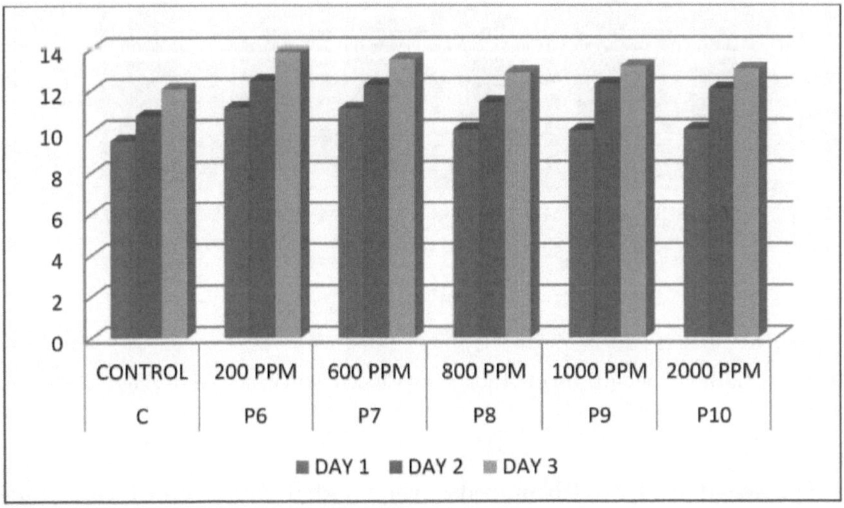

FIGURE 11.31 Shoot growth of *Oryza sativa* added with nanorods of ZnO + 5 ml vine water.

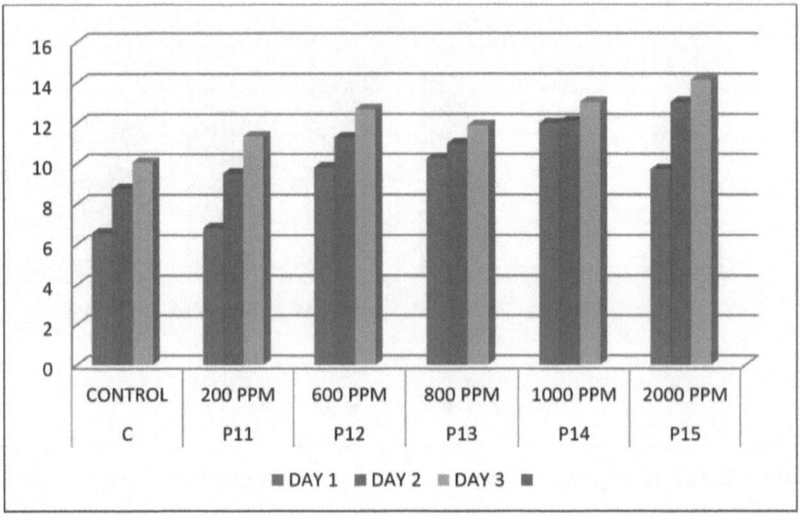

FIGURE 11.32 Shoot growth of *Oryza sativa* added with nanorods of ZnO +15 ml vine water.

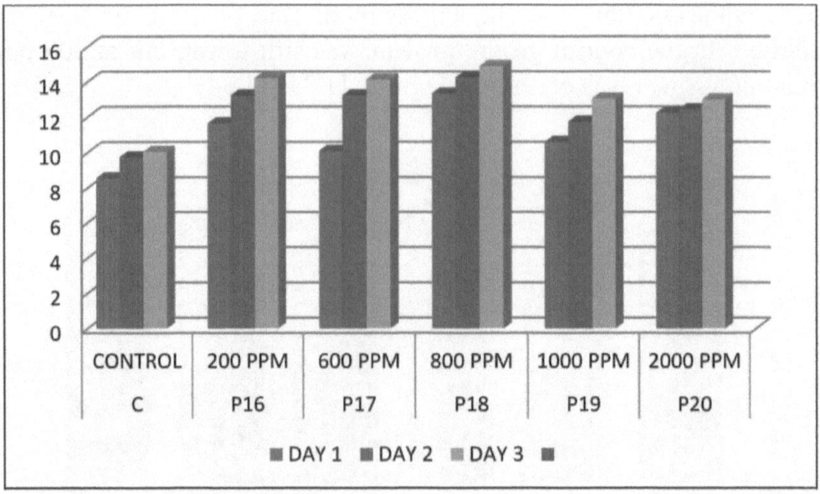

FIGURE 11.33 Shoot growth of *Oryza sativa* added with nanorods of ZnO + 25 ml vine water.

The solution of ZnO Nanorods, when added with 5 ml of vine water, showed a significant increase in root growth. The growth of the root was further enhanced with the addition of ZnO Nanorods treated with

Calycopteris floribunda. The maximum growth of the root was observed in the 1,000 ppm solution (Figure 11.35).

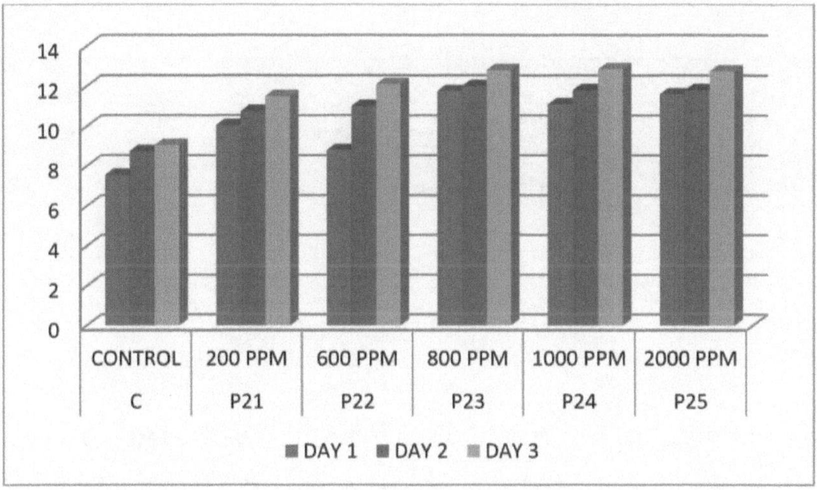

FIGURE 11.34 Shoot growth of *Oryza sativa* added with nanorods of ZnO + 50 ml vine water.

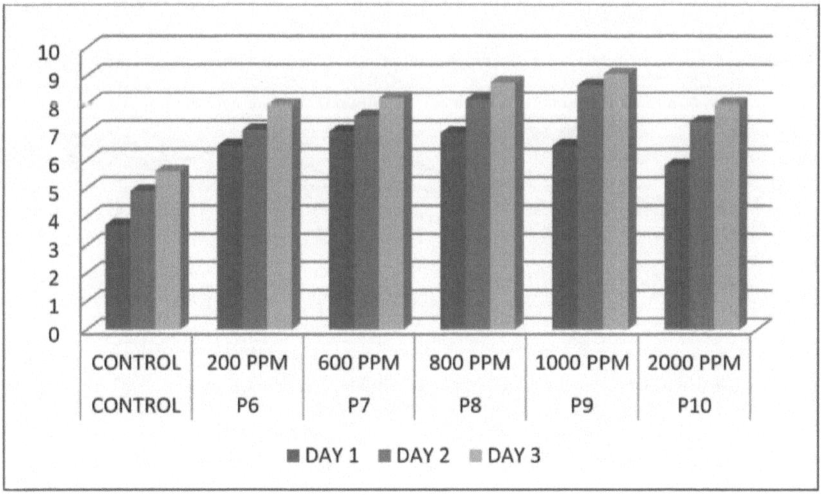

FIGURE 11.35 Root growth of *Oryza sativa* added with nanorods of ZnO + 5 ml vine water.

As depicted in Figures 11.36 and 11.37, it was found that bulk ZnO, chemically synthesized pure ZnO nanorods, and ZnO nanorods treated with *Calycopteris floribunda* (ZnO + 1 ml, 5 ml, 15 ml, 25 ml, 50 ml vine water) of 200 ppm and 1,000 ppm solutions accelerate the growth of shoots in rice plants.

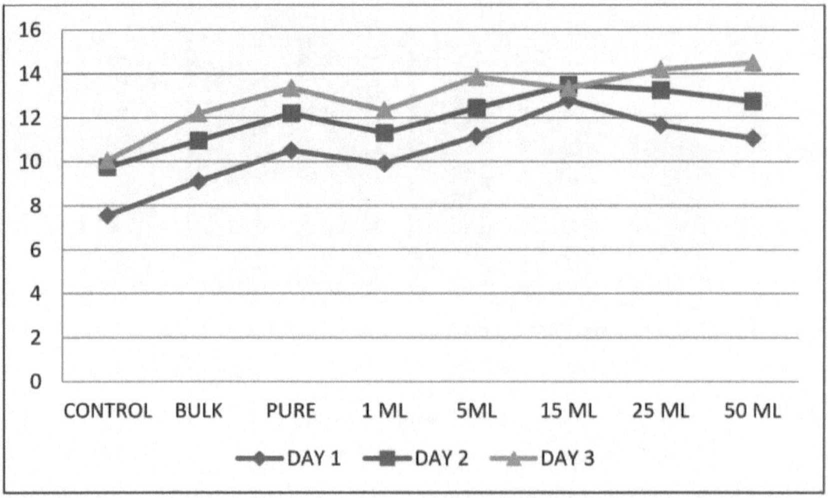

FIGURE 11.36 Shoot growth of *Oryza sativa* added with 200 ppm Bulk ZnO, pure ZnO nanorods, and nanorods of ZnO with different concentrations of vine water.

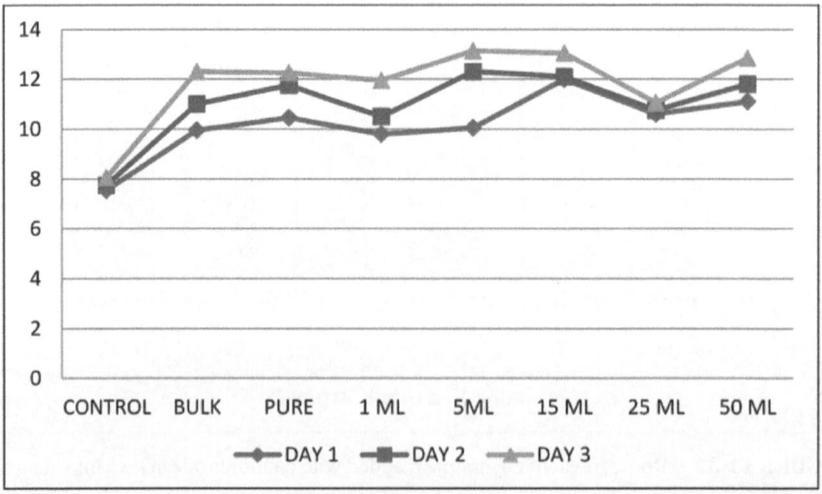

FIGURE 11.37 The shoot growth of *Oryza sativa* was observed when added with 1,000 ppm bulk ZnO, pure ZnO nanorods, and nanorods of ZnO with different concentrations of vine water.

Additionally, the solution of ZnO nanorods added with 15 ml vine water also exhibited significant root growth. The maximum growth of the root was observed in the 1,000 ppm solution (Figure 11.38).

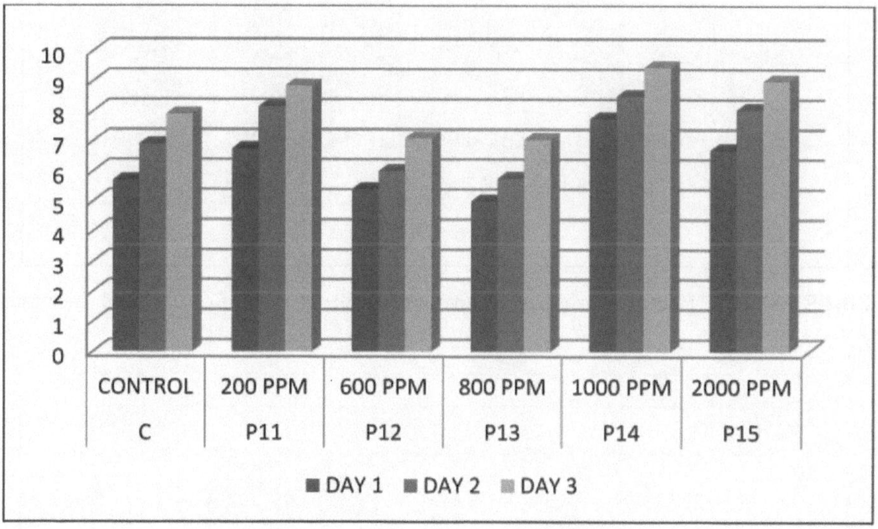

FIGURE 11.38 Root growth of *Oryza sativa* added with nanorods of ZnO +15 ml vine water.

Root elongation was also significant in the Nanorods treated with 25 and 50 ml of vine water from the prepared solutions. Elongation reached its maximum at higher concentrations, such as 1,000 ppm and 2,000 ppm solutions (Figures 11.39 and 11.40).

Results were very significant. There was a gradual increase in the length of the shoot and root with a change in the substrate concentration. It was observed that, compared to the control, bulk ZnO, pure ZnO Nanorods, and ZnO Nanorods treated with *Calycopteris floribunda* supported the elongation of both shoot and root in *Oryza sativa*. However, Nanorods treated with *Calycopteris floribunda* had a more pronounced effect on root and shoot elongation. It is assumed that the antioxidants present in the vine water played a role as inducers and chelators for the growth of the studied plant.

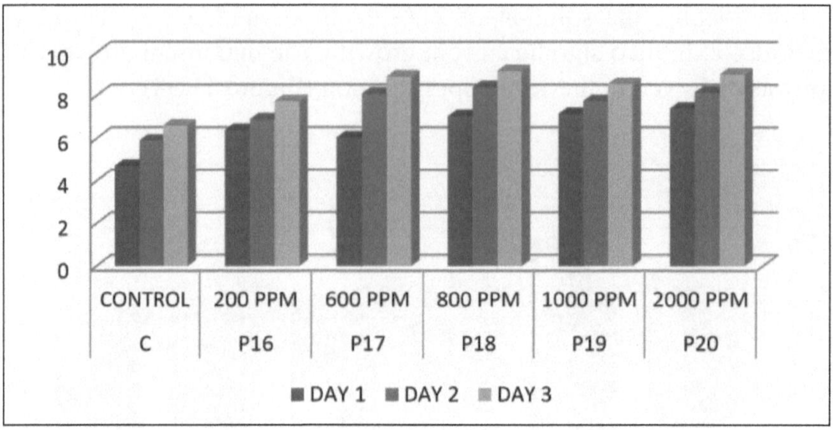

FIGURE 11.39 Root growth of *Oryza sativa* added with nanorods of ZnO + 25 ml vine water.

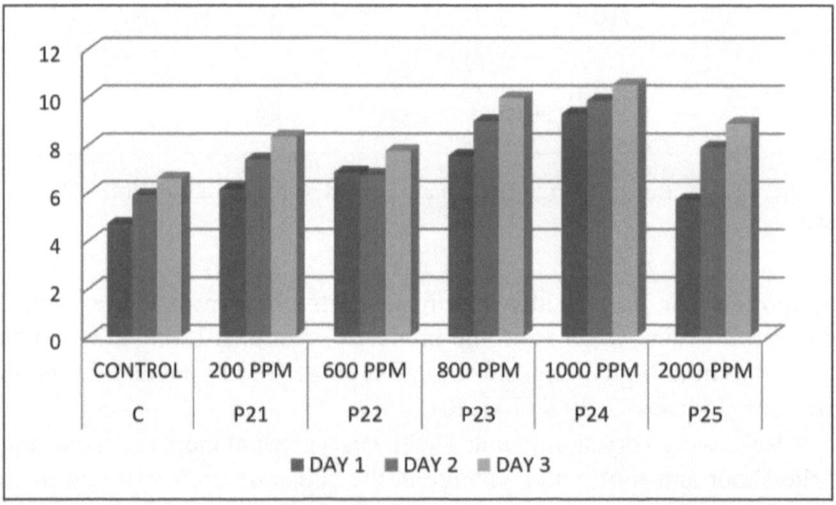

FIGURE 11.40 Root growth of *Oryza sativa* added with nanorods of ZnO + 50 ml vine water.

11.8 CONCLUSION

In the present study, ZnO nanorods were successfully synthesized using both a chemical method and a green approach. The vine water of *Calycopteris floribunda* was utilized in the synthesis of ZnO nanorods to mitigate its toxic effects. Chemical synthesis results in the presence of toxic chemicals like ammonia on the surface of nanorods, which adversely affects their medical and pharmaceutical applications.

SEM images displayed a rod-like morphology for all the prepared samples of ZnO. The synthesized ZnO nanorods ranged from 20 to 40 nm in size. Structural studies conducted by XRD revealed that all prepared ZnO samples exhibited a wurtzite hexagonal structure. The XRD peaks were sharp and intense, and they closely matched the JCPDS values of bulk ZnO. The prepared ZnO nanorods exhibited a highly crystalline structure. Optical properties studied by UV and PL indicated that ZnO nanorods synthesized using the green approach (using vine water of *Calycopteris floribunda*) showed no significant variations compared to those synthesized using the pure chemical method.

Zinc is a micronutrient for plants. To study the beneficial importance of a green approach in the synthesis of ZnO Nanorods, we applied bulk ZnO, chemically synthesized ZnO Nanorods, and ZnO Nanorods synthesized by a green approach to *Oryza sativa* and studied its growth rate. There was a gradual increase in the length of the shoot and root with a change in the substrate concentration. It was observed that, compared to the control, bulk ZnO, pure ZnO Nanorods, and ZnO Nanorods treated with *Calycopteris floribunda* supported the elongation of both shoot and root in *Oryza sativa*. However, Nanorods treated with *Calycopteris floribunda* had a more accelerating effect on root and shoot elongation. It is assumed that the antioxidants present in the vine water played an inducer role and acted as a chelator for the growth of the studied plant. We anticipate that the alkaloids and proteins in the vine water of *Calycopteris floribunda* reduce the toxic effects of ZnO Nanorods and act as a stabilizing agent.

KEYWORDS

- ***Calycopteris floribunda***
- **green synthesis**
- ***Oryza sativa***
- **scanning electron microscope**
- **X-ray diffraction**
- **zinc oxide nanorods**

REFERENCES

Alexandra, N., (2003). Materials and nanotechnology. *J. Franklin Institute, 340*, 263.

Andreas, S., (2011). Energy storage: Batteries take charge. *Nature Nanotechnol., 6*, 262.

Bharat, B., (2004). *Hand Book of Nanotechnology* (3rd edn.). Spring, Germany.

Bogunia-Kubik, K., & Sugisaka, M., (2002). From molecular biology to nanotechnology and nanomedicine. *Biosystems, 65*, 123.

Cakmak, I., (2000). Role of zinc in protecting plant cells from reactive oxygen species. *New Phytologist, 146*, 185–205.

Chopra, R. N., Nayar, S. L., & Chopra, I. C., (1956). *Glossary of Indian Medicinal Plants* (p. 46). CSIR. New Delhi, India.

Daniel, M., (2006). *Nanotechnology Applications to Telecommunications and Networking*. Wiley & Sons, New Jersey.

Garcia-Martinez, J., (2010). *Nanotechnology for the Energy Challenge*. Wiley-VCH DIO: 10.1002/9783527629299.ch1.

Geller, M., Hopfer, F., & Bimberg, D., (2008). Nanostructures for nanoelectronics: No potential for room temperature applications? *Microelectronics Journal, 39*, 302.

Jan, G. K., & Andreas, G., (2002). *Semiconductors for Micro and Nanotechnology: An Introduction for Engineers*. Wiley.

Laura, M., (2003). Commercializing nanotechnology. *Nature Biotechnol., 21*, 1135.

Lloyd, M. S., (2010). Nanotechnology: Molecular robots on the move. *Nature, 465*, 167.

Mark, W., & Jean-Yves, B., (2007). *Environmental Nanotechnology: Applications and Impacts of Nanomaterials*. McGraw-Hill.

Mohamadi, M. R., Mahmoudian, L., Kaji, N., Tokeshi, M., Chuman, H., & Baba, Y., (2006). Nanotechnology for genomics and proteomics. *Nanotoday, 1*, 38.

Mohamed, H., (1998). Nanotechnology-growing in a shrinking world. *Focus on Nanotechnology, 11*, 30.

Pramod, M., Dhoke, S. K., & Khanna, A. S. (2011). Effect of nano ZnO particle suspension on growth of mung (*Vigna radiate*) and gram (*Cicer arietinum*) seedling using plant agar method. *Journal of Nanotechnology, 2011*.

Prasad, T. N. V. K. V., et al., (2012). Effect of nanoscale zinc oxide particles on the germination, growth, and yield of peanut. *Journal of Plant Nutrition, 35*, 905927.

Rainer, W., (2008). *Nanotechnology: Information Technology* (Vol. 3, No. 1). Wiley-VCH.

Sahoo, S. K., Parveen, S., & Panda, J. J., (2007). The present future of nanotechnology in human health care. *Nanomedicine, 3*, 20.

Shah, V., & Belozerova, I., (2009). Influence of metal nanoparticles on the soil microbial community and germination of lettuce seeds. *Water, Air and Soil Pollution, 197*, 143–148.

Shindo, D., & Hiraga, K., (1998). *High-Resolution Electron Microscopy for Material Science*. Springer, New York.

Soosen, S. M., Lekshmi, B., & George, K. C., (2009). *Optical Properties of ZnO Nanoparticles* (Vol. 5, pp. 7–65). ISSN: 0973-7464 Vol.XVI: No. 1 & 2 SB Academic Review.

Whitesides, G. M., (2003). The "right" size in nanotechnology. *Nat. Biotechnol., 21*, 1161–1165.

Index

D

E

H

I

T